Stepping-stones to improve upon functioning of participatory agricultural extension programmes

Stepping-stones to improve upon functioning of participatory agricultural extension programmes

Farmer Field Schools in Uganda

Prossy Isubikalu

Wageningen Academic
Publishers

This book was originally written as PhD thesis.

ISBN: 978-90-8686-021-0

First published, 2007

Wageningen Academic Publishers
The Netherlands, 2007

Table of contents

List of tables

List of figures

List of boxes

List of acronyms

A2N	Africa 2000 Network
AESA	Agro Ecosystem Analysis
ARDC	Agricultural Research Development Center
ASPPA	Abuket Sweet potato Producers and Processors Association
AT	Appropriate Technology
CAO	Chief Administrative Officer
CBO	Community Based Organisation
CIAT	International Center for Tropical Agriculture
CIP	International Potato Center
COARD	Client Oriented Agricultural Research and Dissemination
CPR	Crop Protection Research programme
DANIDA	Danish International Development Agency
DAO	District Agricultural Officer
DAP	Di-ammonium Phosphate
DEC	District Extension Coordinator
DFID	Department for International Development
FAO	United Nations Food and Agriculture Organisation
FDC	Forum for Democratic Change
FEWs	Field Extension Workers
FFS	Farmer Field School(s)
FOSEM	Food Security and Markets for Small Holders
FYM	Farm Yard Manure
HYV	High Yield Varieties
i@mak.com	Innovations at Makerere Committee for the Community
ICM	Integrated Crop Management
ICRAF	International Center for Research in Agro-Forestry
IDM	Integrated Disease Management
IFAD	International Fund for Agricultural Development
INSPIRE	Integrated Soil Productivity Improvement Initiative through Research and Education
IPM	Integrated Pest Management
IPPHM	Integrated Production and Post Harvest Management
IPPM	Integrated Production and Pest Management
KARI	Kawanda Agricultural Research Institute
LCIII	Local Council at sub-county level
LCV	Local Council at district level
MAAIF	Ministry of Agriculture, Animal, Industry and Fisheries
MAK	Makerere University
NAADS	National Agricultural Advisory Services
NAARI	Namulonge Agricultural and Animal Research Institute

NARI	National Agricultural Research Institute
NARO	National Agriculture Research Organisation
NARS	National Agriculture Research System
NGOs	Non-Government Organisation(s)
NRI	Natural Resources Institute
NRM	National Resistance Movement
OFSP	Orange Fleshed Sweet Potato
PEAP	Poverty Eradication Action Plan
PMA	Plan for Modernisation of Agriculture
PRSPs	Poverty Reduction Strategic Papers
SAARI	Serere Agricultural and Animal Research Institute
SG2000	Sasakawa Global 2000
SOCADIDO	Soroti Catholic Diocese Integrated Development Organisation
ISPI	Integrated Soil Productivity Improvement
SPUH	Safe Pesticide Use and Handling
T&V	Training and Visit
UNSPPA	Uganda National Seed Potato Producers Association
VEW	Village Extension Worker

Acknowledgements

This work, before you, is what it is because of various processes and people. It is not only investigating a social process but developing it to this level was in itself a social process. The goal to attain the highest education level started as an innocent idea in my mind when I was still battling with my primary education. At that time, I did not even know that there was something called a Ph.D. neither did I have a clue about the possibility of pursuing the same. As I walked up the education ladder at University the goal got fine tuned with encouragement from many people who made different contributions, making me what I am. I am therefore indebted to several individuals and organizations whose contributions in one way or the other have shaped me during this PhD path.

Sincere appreciation and gratefulness to the Rockefeller Foundation for fully supporting this study through the Participatory Approaches and Up-scaling (PAU) programme. Thanks to the PAU programme and the Technology and Agrarian Department (TAD) chair group for support and guidance throughout the course of this study. Special thanks to Conny Almekinders the coordinator of PAU for her support in making our stay in Wageningen comfortable. Your care, effort to listen and attend to individual needs of the PAU group is appreciated. Inge Ruisch, you were very helpful in ensuring I travelled safely to and fro Wageningen and Uganda. The smiles and jokes we always shared remain new and pleasant. Thanks to my employer, Makerere University, for granting me a study leave that made it possible to fully engage in this study and to my colleagues in the Faculty of Agriculture especially Department of Agricultural Extension Education for the encouragement.

Professor Adipala Ekwamu, I owe you a lot for you have been one strong pillar in my postgraduate academic life. You are a parent who kept me on my toes to make sure I harvest the best at any moment. You rekindled the idea of doing a PhD and went an extra mile to secure an opportunity to make the dream a reality through the Rockefeller Foundation. Your words and performance orientation remain a strong inspiration.

I remain challenged by the invaluable guidance and support that I received from my promoter, Prof. Paul Richards and co-promoter Dr. Harro Maat. The questions and suggestions from Paul at first sounded impossible, some appeared simple but very valuable. Those suggestions broadened my worldview and thinking and added to the quality of this thesis. Harro, you always created time for me and comforted me in many situations when the going got tough. I am very lucky and grateful to have worked with you and will always remember you.

Being away from home is stressing and affects one's output. I thank my colleagues in PAU for their support that relaxed my life while pursuing this study. Special thanks to Paul Kibwika, Geoffrey Kamau, Francisco Guevara Paco, Chris Opondo, Bernard Kamanga and Budsara, you always encouraged me to work. Doris Kakuru, James Semuwemba, Diana Akullo, Richard Mugambe, Enoch Kikulwe, Fred Bagamba, and Fiona Mutekanga. Thank you for being friends. It was a joy to meet and interact with you. The smiles and jokes we cracked served as fuel in me. You made me feel at home and enjoy my stay in Wageningen. Interacting with you made a big difference in my social life. Many thanks.

Without help from some FFS coordinators, programme assistants, extension workers and farmers in the study districts (Soroti, Tororo, Busia, Kiboga, Iganga, Mukono, Kumi), it would not have been possible to get data that resulted into this book. James Okoth, Sam Namanda, Hire Jenifer, John Ereng, Opio Geoffrey you were very instrumental in linking me up to the different people in the field. Thanks a lot. My sincere thanks to all the institutions that allowed me to carry out my study on their FFS projects. Crop Science Department - Makerere University, Africa 2000 network, CIP and FAO.

Lastly, I am indeed indebted to my mum, brothers and sisters for their overwhelming support and encouragement. Jolly, you are unique. I do not know how to thank you. You accepted to stay at my home for almost a year and postponed your University studies to care for my boy Wycliff Nandigobe whom I left at a tender age for my PhD studies in Netherlands. That was very kind of you. Fred Nandigobe, you did not only play your role as a father but also cared for Wycliff as a mother. That kept me firm in pursuing my studies to the end. May God reward you abundantly. Wycliff, my beloved son, I missed attending to you as a mother for all the period I was carrying out this study. I wish to dedicate this thesis to you my son. Because with God everything is possible, I am hopeful for a better future, with God's Grace.

Prossy Isubikalu, April 2007

CHAPTER ONE

Participation and poverty reduction: strategy in effecting agricultural extension programmes

1.1 Introduction

This thesis is about participation in people-centered agricultural extension programmes or systems. It describes the interactions involved, and their outcomes. The research is based on a technographic approach. Technography (discussed further below) can be defined as the systematic description and analysis of the interaction of human agents, tools, techniques and technical processes, i.e. it is the study of instrumentality within the broader field of ethnography. It is applied, here to document what actually happens when poor farming people in an African country (Uganda) become involved in shaping agricultural technology development through participatory extension.

Why has participation become a leading concept among development agencies? Despite extensive aid investments over recent decades poverty has remained stubbornly high, and has even increased, in many of the poorest countries. By way of explanation, it has been suggested that a major factor is the persistent detachment of development actors, at national and international levels, from realities and priorities of communities at grass root level. Participation of the poor in defining their own goals for development and working towards feasible outcomes is supposed to overcome this problem. The shift is driven by policy instruments - namely, the requirement of impoverished countries to prepare poverty reduction strategies (PRSPs) and also to develop strategies for delivery of Millennium Development Goals (MDGs) as triggers for debt relief and continued development assistance.

Through the Poverty Reduction Strategy Papers (PRSPs) framework participation is used as a method to link up communities at the lower levels with their governments at higher levels. This link is intended to guide resources more effectively towards the priorities of the people at the community level. The PRSP approach requires poor country governments to develop and forward their own individual long term development plans - i.e. the PRSP - to the International Monetary Fund (IMF) and World Bank for financial support. Presentation of a PRSP is treated as a pre-requisite for full debt reduction, a factor motivating many deeply indebted countries to adopt the approach.

The MDG approach has a rather different basis. The millennium treaty was signed by countries in the United Nations general assembly, and binds all parties to work towards meeting a number of specified poverty reduction targets by the year 2015. Poor countries need to define viable strategies, and rich countries are obligated to provide assistance where these targets cannot be met upon the basis of local resources alone. Progress towards targets is monitored, and countries falling behind are then supposed to be assisted to "raise their game". Meeting MDGs requires mass mobilization of local human and social "capital" as well as outside assistance. Participation is seen as a central aspect of this mass mobilization.

Participation of civil society, articulation of voices of the poor and the reduction of poverty through mass mobilization, as implied in PRSPs and MDG frameworks constitute an agenda for transformation with apparent moral authority. Participation is a normative requirement - an aspect of democratisation and the creation of an open, accountable relationship between rulers and the ruled. Participation and poverty reduction, however, may also be used as words for fine-tuning what amounts to direction "from above" in development policies (Cornwall and Brock, 2005: 1044). In other words, they may be deployed to lend legitimacy to donor interventions motivated by international political considerations (e.g. the need to solidify certain kinds of client relationships among poor countries for the pursuit of global political objectives). In such cases, participation is a term of theory (specifically of international relations) rather than a practice-oriented concept. For the most part, in this thesis, we treat participation as a practice-oriented notion, motivated by a true concern for democracy, and leave aside the cynical interpretation.

Within the practice-oriented framework we can identify a number of basic issues concerning participation. First is that there is a problem in actually getting people to participate, and this has been one of the reasons why participation has often been better supported in principle than in practice (Simmons and Birchall, 2005). Second, mobilisation methods for getting the community to take part in development projects and to develop high levels of commitment also need to be addressed. Third, there should be clarity about the objective of calling upon the community to participate: is participation a means to an end or is it an end in itself? Finally, in general terms, it is important to recognise that participation as a basis for any project or activity may not necessarily make a difference or lead to meaningful change in policies if a comprehensive look at what is actually done in the name of participation is not followed up with careful critical assessment of what it amounts to in terms of realities on the ground (the technographic aspect).

What is written, and said in principle, often differs from what is done in practice. For instance all countries emphasize the importance of economic growth for poverty reduction, but some PRSPs reviewed have not been sufficiently pro-poor (see Booth, 2003; UNDP, 2001a). Some reasons for insufficient pro-poor status include:

- focus on reducing income poverty while tending to to ignore structural determinants of poverty;
- thinking for communities about what works instead of involving them in identifying issues that need attention for their own good;
- not listening to voices from grassroots;
- Assuming poor communities are homogenous and cooperative rather than (as is often the case) riven by inequality and conflict.

In the developing world, the largest population is in the rural areas where agriculture is one of the main occupations and sources of livelihood. Ignoring farmers (a term here used very broadly to encompass not only "land owners" but a broad group of farm household dependents from whom much of the actual farm labour is derived - typically wives and children [especially girls] in the Ugandan situation) means not caring about a key focus of the poverty problem. Yet, farmers are important agents in any innovation system geared towards improving the socioeconomic conditions of rural communities and the development of relevant science and technology (see Van Mele *et al.*, 2005). Inadequate technologies from researchers and blue-print blanket recommendations from extension agencies contribute to an undesirable situation in which farmers are seen as recipients and not partners in (technology) development, supporting a one-size fits all approach. It is an effort to get beyond the well-recognised weaknesses of this top-down approach (sometimes termed ToT - transfer of technology) that interest has been shown in developing participatory approaches to technology generation and technology extension for impoverished farmers.

The present study focuses on interactions between technology and society in an African agrarian context in which policy is driven by PRSP and MDG considerations. It is a "technography" of a specific kind of participatory innovation system, the Farmer Field School model (henceforth FFS), aimed at improving upon existing agricultural technology generation and extension systems directed towards supporting the production and productivity of poor Ugandan farmers under conditions of "market failure". "Market failure" here refers to the situation where farmers are so poor that they cannot buy available and necessary technology services, and where (in any case) these services fail to meet the needs and interests of the rural poor due to the predominance of the ToT approach. Service supply, in short, does not meet community demand in poor communities. Commercial pressure is too weak to "induce" the right kind of technological innovation and farmers have to be stimulated to participate, in order to adopt, adapt, and invent new solutions to technological constraints. These solutions need to take into account family survival as well as market opportunity (because poor farmers have limited time and resources to invest in innovation activity with uncertain outcomes). The situation is made even more challenging due to government detachment from rural realities, corruption, and the dislocations associated with war over many years (now being addressed), which led to chronic underinvestment in basic infrastructural requirements (roads, health, basic education, market infrastructure).

1.2 Agriculture as an important entry point in PRSP

Under PRSP, agriculture is viewed as one entry point in ensuring poverty reduction among rural African populations. Agriculture has not figured high on the donor agenda in recent decades. Rural poverty has meanwhile proved stubbornly persistent, and policy makers at global and national level have begun to realise the need to address poverty directly in the rural and farming communities where it is most prevalent (Hazell and von Braun, 2006). If the rural poor are the target, then rural development and small-scale agriculture need to figure prominently on the poverty reduction agenda. To ensure that the poor benefit from economic growth many PRSPs propose a development strategy that prioritises agricultural development and stresses spending on poverty reduction (Gottschalk, 2005: 424). For sustainable development and agricultural improvement agricultural extension services to encourage cooperation and collective work are advocated as main entry points in a number of PRSPs. Improved agricultural extension is about working with a multitude of actors in effort to come up with an agro-technical system commensurate with prevailing local realities. A leading idea is to boost participation of communities, in the rural areas, by getting them to take an active role in shaping development and poverty-alleviation decision making processes. Extension based on the idea of instructing farmers in new methods is replaced by the idea of interaction around problem definition and problem solving. This participatory emphasis in extension aligns with the larger understanding, under PRSP and MDG approaches to poverty alleviation, that participation is a principle by which the sustainability of a national development strategy can be assessed: in other words, progress towards poverty eradication (the MDG agenda) is measured in terms of how well the short and long term needs of disadvantaged and marginalised groups are integrated within economic policy (Cherp *et al.*, 2004). The PRSP/MDG approach requires participation of rural communities and other stakeholders to be institutionalised within national policy frameworks in poor countries. This puts an onus on both country and community to attain improved ownership of policy design and implementation processes adapted to local realities. However, it has been observed that participation in PRSP development is often tightly controlled from above (Sanchez and Cash, 2003; Christian Aid, 2001). It is questionable how much in practice stakeholders from among the poor are actually actively involved in influencing policy and programs with their priorities. Inadequate participation of societal actors, and lack of transparency, makes genuine participation in PRSP a challenge (see Cheru, 2006). This thesis aims to assess how well this challenge is being met, through the specific activity of FFS.

In Uganda, the country central in this thesis, an institutional window, in the form of the Poverty Participatory Assessment Project (henceforth PEAP), created space for government to interface with civil groups (UPPAR, 2000; McGee, 2002: 70). The representatives of (so-called) civil society, however, mainly work at a national level, and tend to lack close contact with realities in the rural areas. The smaller rural self-help groups directly feeling the pain of poverty at the grassroots are often left out of the consultation process! This directs attention to a key question underlying the detailed analysis in this thesis. What light can be thrown on the problem of incorporating the voices of the poor in poverty-alleviation strategy by examining participation of people at the grass roots in Farmer Field Schools? Does this modality of participation offer real prospects to incorporate the voices of the poorest in civil society deliberations about national development policy, or is the process stage managed in such a way that the agenda of the well-connected is reproduced? In short, this thesis is concerned with whether FFS is useful in enabling truth to speak to power, or are the participating poor simply puppets in a game of speaking power to power.

1.2.1 Uganda's commitment to rural development: the PEAP

With a population of about 25 million people, of which 86% are in the rural area and 77% actively engaged in agriculture (UBOS 2002)[1], agricultural growth is seen as critical for poverty reduction and rural development. A majority of farming communities are engaged in semi-subsistence agriculture (i.e. mainly for own and local consumption) and tend to be left on their own because they seem not to contribute directly to economic growth via exports. Although poverty levels decreased from 54% to 36% between 1992 and 2000 (Deininger and Okidi, 2003) engagement in crop agriculture remained the most important contributor to increased poverty after 2000 (Kappel *et al.*, 2005). The predominantly subsistence farming community in Uganda is more engaged in food crop agriculture than cash cropping or livestock agriculture. A ready local market for food crops, and their dual utility (that which is not sold can be eaten) explains why farmers are more engaged in food crop agriculture. Food crop agriculture, however, pays less and keeps the majority of the communities in the rural areas in poverty: one reason is that during bumper harvests nearly everyone has produce (food) and marketing it becomes a problem.

In response to the commitment to alleviate poverty, the government of Uganda embarked on a number of initiatives and strategies emphasising rural development. A Poverty Eradication Action Plan (PEAP) was designed and developed in 1997 with donor funds. The main objective of the initiative was to reduce the proportion of the population living in absolute poverty to 10% by the year 2017[2] (MAAIF and MFDEP, 2000). Uganda is also actively directing resources through the Poverty Action Fund (PAF) to social development, with particular focus on rural transformation and modernisation of agriculture. PEAP was/is Uganda's national development framework and medium planning tool as well as the guiding formulation of government policy and implementation of programs through a sector-wide approach and decentralisation for PRSP (MFPED, 2004:13). Under PEAP, priority action areas identified for effective poverty eradication included primary health care, roads, primary education, rural water, and agriculture. To address poverty through agriculture, a plan to modernise agriculture was developed. This plan was not limited to agriculture, as the title might otherwise seem to suggest. It was broader in scope, and involved all sectors related to or influenced by agriculture.

1.2.2 Uganda's plan for the modernisation of agriculture

In recognition of the multiple factors impeding the attainment of rural and agricultural development, the government of Uganda formulated a comprehensive Plan for Modernisation of Agriculture (PMA) that was aligned with PEAP goals. The focus of the PMA is to transform agriculture from a subsistence to a commercial orientation. The PMA was designed (by

---

[1] According to the most recent population census conducted in September 2002, total population was 24 million. However with an annual growth rate of 3.3 the estimated population in 2005 is 26.7 million.

[2] The Millennium Development Goals are targeted on reducing the proportion living in extreme poverty and hunger by 2015.

government and donors) around seven pillars specifically to address factors that undermine agricultural productivity: poor husbandry practices, low use of and access to improved inputs, limited access to technical advice, poor access to credit, poor transport, communication and marketing infrastructures, and insecure land tenure and user rights were identified as major constraints undermining agricultural production and thereby promoting poverty in the country. The constraints (above) were categorized into broad areas forming PMA pillars: agriculture research and technology development, delivery of agricultural advisory services, rural finance, promotion of agricultural marketing and agro-processing, agricultural education, sustainable natural resource management and use, and supportive infrastructure. Gender and HIV/AIDS were treated as cross cutting issues.

All projects, as a matter of policy, fitted within the PEAP/PMA framework. Guidelines and vetting committees - the PMA steering committee and development committees (PMA SC & MFPED DC 2003) - were put in place to ensure harmony of projects within the policy framework. A recent PMA evaluation exercise (Oxford Policy Management, 2005) revealed that farmers perceived poverty to be on increase. Low income, limited human development and limited community empowerment were the main aspects of poverty thereby revealed (PEAP, 2004). Empowerment largely referred to confidence and ability to define goals and work towards achievement of such goals. This implied a major effort to remove the kinds of subordination, especially at the top, where policy had long been developed without involvement at the grass-roots level.

Rural farming communities, and women in particular, are commonly victims of power relations that make them vulnerable to social, economic and political shocks. Inadequate participation of communities in contributing to decisions that affect them remains a very important constraint. Actually, people themselves perceive poverty mainly in terms of an "inability to satisfy a range of basic human needs that stems from powerlessness, social exclusion, ignorance and lack of knowledge, as well as shortage of material resources" (UPPAP 2002: 10). Social analysts claim powerlessness, exclusion, and lack of self expression as the major contributions to increasing poverty among rural communities (Narayan, 2000). Lack of participation inhibits lack of confidence. It is on this basis that participatory methods have been strongly pushed in the context of the changed policy environment for development interventions in Uganda.

1.3 Re-organisation or transformation of agricultural extension in Uganda

Efforts to modernise agriculture and transform agro-technology depend quite centrally on agricultural extension. Agricultural extension is a bridge between technology users (farmers) and technology developers (researchers). The bridging role applies to both formal and informal settings (i.e. extension agents work with farmers in the market sector and also with those affected by market failure, and dependent on their own subsistence efforts). Approaches can be both direct (e.g. offering farmers direct advice on new products and services) or indirect (working on community dynamics in such a way as to create greater interest by farmers in acquiring new inputs and skills for themselves). Either way, it is a basic assumption that greater efficiency, effectiveness and responsiveness on the part of extension services will translate into better agricultural performance, with the greater likelihood that rural development will be attained. In search of an efficient agricultural extension system to enhance people's ability to make appropriate use of opportunities around the extension system in Uganda has varied its approach over time. The changes can be traced from a time when Uganda was still a colony under British administration (see Opio-Odong, 1989). Based on where emphasis was put during the transformation of agricultural extension from 1898 to 2002, the evolution of Uganda extension can be categorised into a number of distinct time periods. Semana (2002) distinguishes eight phases: 1898-1907, the early colonial period; 1920-1956, extension service through chiefs; 1956-1963, extension through progressive farmers; 1964-1972, extension methods phase; 1972-1980, non-directional or dormant phase; 1981-1991, recovery period; 1992-1997, agricultural extension reform; and 1998-2002, crossroad, dilemma and future solutions.

The once purely regulatory system that was reinforced with help of chiefs during colonial times changed to a more advisory and participatory approach. Originally, agricultural extension was exclusively aimed at boosting production of colonial cash crops (viz. coffee, cotton, tea, cocoa, tobacco and rubber) to generate income and foreign exchange for the government. Cash crops were seen as a major source of capital to be re-invested in industrial development (a presumed universal future). Focus on the traditional cash crops (Semana, 1989) implied ignoring food crops, and therefore extension did not cater for food security. Later, specific subsistence food crops (like beans, maize and bananas) were brought on board, given their status as the main food crops for the majority of the population, especially in the central region (Sibyetekerwa, 1989; Opio-Odong, 1992). But the extension approach remained commodity based, and centred on focal persons (termed progressive, contact or model farmers) expected to disseminate knowledge about commodities to other farmers.

Development was then supposed to result from a trickle-down effect. The trickle-down philosophy did not work, however, because contact farmers did not disseminate the recommended practices to other farmers as expected (Semana, 2002). To include as many farmers as possible the extension approach was changed to one that encouraged participation of the entire community at village level. This was known as the Village Level Participatory Approach (VLPA). VLPA was designed as a community development initiative to reinforce bottom-up planning and implementation processes. Under the Agricultural Extension Program (AEP) implemented in 1992, a unified extension approach was initiated. This later involved the introduction of the Training and Visit (T&V) methodology, to ensure propagation of recommended practices in a logical and systematic manner (Midland Consulting group, 1997; Mubiru and Ojacor, 2001). Although T&V improved on the effectiveness of existing extension personnel through in-service training, it still embodied a top-down (instructional and ToT) approach, rather than emphasising discovery-based learning. This proved inappropriate to address farmers' very varied realities on the ground.

Despite all the changes in extension aimed at providing information, knowledge and skills, useful to and compatible with resource poor farmers living in fragile and ever changing socio-economic and biophysical environments, farmers continue to live with the same problems. Many reasons account for this situation. According to Aben *et al.* (2002), programs and activities implemented did not always represent key priority enterprises of a majority of the poor. The focus was often more on cash crops than food crops, where a majority of the rural community was engaged in food crop production. Lack of clearly defined and focused agricultural policy, insufficient training of extension staff in extension methodology, ambiguous missions and objectives, unsuitable organisational structure and administrative arrangements undermining staff morale and misappropriation of limited funds have been listed (Semana *et al.,* 1989) as the main attributes of extension failure, in spite of regular changes from one approach to another. Passivity at the community level and a tendency to treat all farmers (and their contexts and needs) as homogenous are additional invisible contributions to the failure of conventional extension programmes.

It is against the above background that participatory, demand-driven, client-oriented, and farmer-led agricultural extension systems, with emphasis on targeting the poor and women, have been advocated. These categories of rural people are marginalized, yet their contribution to the economy is greater than often realized. This neglect stems from the fact that most of them are not directly engaged in growing traditional cash crops, like coffee, that earn foreign currency for the government. But they do contribute to the food supply, including the food supply of those who farm, market and process crops like coffee. Provision and availability of food for the entire population is an essential task that has been largely left to the labour and efforts of women and the poor who dominate the rural and agricultural sector. Therefore, any policy that affects agriculture automatically affects them and likewise any policy that affects women or rural farming communities is likely to affect agriculture and the economy more largely.

The National Agricultural Advisory Services (NAADS), a key component of Ugandan agricultural extension under PMA, was developed as a new system to replace the old and non-responsive conventional extension system. NAADS in principle emphasises participation of communities in decisions about enterprises that fit their needs at specific local levels. It was designed to increase farmers' access to improved knowledge, technologies and information. The NAADS programme, implemented from 2001, is grounded in the government's overarching policies of agricultural modernization, poverty eradication, decentralization, privatization, and increased participation of the people in decision-making (Nahdy, 2004). In re-organising agricultural extension services in Uganda, all agricultural extension related projects are required to fit within NAADS framework.

During the last decade the Food & Agriculture Organization of the United Nations (FAO) has been a key player in developing the Farmer Field School (FFS) model to address some of the weaknesses of "top-down" extension practice. The FFS approach responds to the call to strength farmer knowledge and skill not through instruction and supervision but through active experimentation and group learning. This responsiveness to specific local needs and active involvement of farmers in technology development complies with the PEAP/PMA policy framework of encouraging greater participation by the poor in addressing their own problems. International donors were willing to support the government of Uganda in developing NAADS as a client-oriented extension system, and FFS has been incorporated as a mini programme to help in this re-orientation. However, as we will see, FFS applied to the problems of mixed cropping in Uganda has evolved quite a long way from its origins as an approach to pest control problems in (rice-based) mono crop farming systems in Asia.

This leads directly to a statement of the broad research question addressed in this thesis. *In what ways does the FFS model re-organise the agricultural extension system in Uganda and serve to improve the ability of that system to address farming realities?* Building confidence and analytical and decision making skills among farmers are said to be among the empowering ingredients acquired through the learning process in FFS. The FFS model envisages that the more knowledgeable farmers become, the more confidently and effectively they will make key decisions. Thus FFS has a bearing on poverty reduction not only via improving agricultural output but through the increased capacity of impoverished farmers to make skilful decisions. However, this empowerment objective depends on how the whole process of implementation and operation is handled. Before venturing into how FFS was introduced and used as a mini program to reformulate the agricultural extension system in Uganda it is worthwhile to take account of how the model assumed its present form.

1.4 The origin and development of Farmer Field Schools

The Farmer Field School (FFS) model is linked, historically, with Integrated Pest Management (IPM). It was developed as a way of introducing rice farmers in South-east Asia in the late 80's to more appropriate and ecologically sustainable agricultural practices resulting in reduced pesticide use. It is worthwhile to briefly venture into the evolution of IPM, in order to understand the task of evaluating its performance in very different operational circumstances in Uganda.

1.4.1 Development and evolution of IPM

IPM is a crop protection concept. As an ecological-based approach to pest control strategy, IPM practice adjusts to changes in pest threat. It is a rolling adjustment approach, since pest problems change over time as the pests themselves evolve. Early efforts were focused crop protection problems thrown up by the Green Revolution in rice. The Green Revolution developed high yielding crop varieties (HYV). Use of the HYV boosted agricultural production. But HYVs were susceptible to pest attack. Mono-cropping of densely planted new rices selected for certain kinds of pests (e.g. Brown Leaf Hoppers) which underwent population explosions. This situation forced farmers to apply high levels of pesticides as a control measure. These pesticides were expensive and had a number of deleterious environmental effects. A second generation of pest resistant rice varieties was developed to minimize the use of pesticides. Development of pest resistant varieties, however, did not stop farmers from using pesticides. Breeding resistance into plants might only work for specific pests regarded as important at that time, and for crop types deemed as being in the national interest (e.g. suited to export markets). Resistance breeding was not a strategy for all crops, and farmers continued to use pesticides for pest problems on these other crops.

Increasing pesticide use raised scientific concern about effects on the environment and human health. There was a general desire to ensure reduced or judicious use of pesticides. The concepts of economic threshold (ET) and economic injury level (EIL) (Stern, 1973; Pedigo et al., 1986) were developed in the 60s, as tools of pest control strategy based on arriving at the minimum pest density to justify pesticide spraying. But these too focused on one crop and the important pests. ET required some level of observations and scouting for the pests. ET and/or EIL were not very favorable to contexts in which farmers faced a complex of pests. In order to manage more than one pest and minimize pesticide use it seemed more promising to use a combination of cultural and biological methods. This notion was brought to fruition in the concept of IPM.

As IPM evolved, farmers and the farming system became increasingly important as focuses of technological innovation. The role of farmers in managing pest problems became more evident. This can be seen in the ways in which IPM was defined from the 70's to date. In earlier formulations, the concept of IPM was more one of integrated control than integrated management, and the role of farmers and their practices in managing or influencing pest population dynamics tended to be downplayed or ignored. This first generation definition of IPM focused on integrated control as a pest management system that in taking account of the context of associated environment and the population dynamics of pest species uses all suitable techniques and methods in as compatible a manner as possible and maintains the pest population at levels below those causing economic injury (FAO, 1967). This targeted more pests and encouraged combinations of host resistance, cultural and selective chemical control, while targeting more pests than was typical of ET and EIL.

In the 80s, the US Council for Agricultural Science and Technology (1982) modified its definition of IPM - pushing it to a higher system level - as the use of two or more tactics in a compatible manner to maintain the population of one or more pests at acceptable levels in the production of food and fiber crops while providing protection against hazards to humans, domestic animals, plants and the environment. At this level, the role of human beings in influencing pest management and the need for a holistic crop management strategy was catered for.

From the 90s to date, there have been efforts to integrate technical and social (techno-socio) sciences in IPM development as well as in its implementation. There are various definitions, as perceived by different people. At present, social, economic, and political variables, and entire farming systems, are aspects taken into consideration while thinking through IPM. Indeed, pest problems arise through interactions between human and natural systems at different levels. Norton and Mumford (1993) described these systems in a hierarchical order from pest system, crop system, farming system, village system, provincial system and national system. Interactions are within and between the systems. In the techno-socio approach to IPM it becomes clear that local conditions other than prescriptions of recommended scientific practices determine the most appropriate pest management technology. Consequently emphasis is put on location-specific sustainable agricultural and pest management practices for effective management of the agro-ecosystem. Implementation of the techno-social type of IPM requires an appropriate design, delivery and training method. FFS meets this need. FFS was first used in South-eastern Asia as a dynamic approach taking into consideration the socio-economic and political conditions, among other factors, that influenced farmers in taking up (IPM) technology.

1.4.2 Farmer Field Schools in South-East Asia

The Green Revolution seed technologies (HYV) combined with scheduled fertilizer and pesticide application enormously boosted rice production in Asia. Excessive and indiscriminate use of pesticides, however, took off the shine from this technological success. In an effort to save the situation, the government of Indonesia banned 57 broad spectrum pesticides, and institutionalized IPM as a national pest control strategy (Winarto 2004). Backed by adequate experience in crop protection and rice cultivation practices, FAO established an IPM programme in seven Asian countries in 1979. As a crash program, IPM was first introduced in Indonesia through the then existing Training and Visit (T&V) extension system, where technology packages were prescribed to the (rice) farmers with a key message of spraying less (van der Fliert, 1993). Although this extension message restored part of the ecological balance (Kenmore, 1991), it was not followed by many farmers and was not sustainable. Farmers were locked in rigid schedules recommended by extension through T&V and remained unshakeably convinced that high input (pesticides) led to high output.

Something was needed to make farmers realize the effects of high pesticide use on the ecosystem. Integrating production and productivity technologies/practices with sustainable IPM practices aimed to meet this need. This meant a new model of extension, since the conventional T&V approach was not appropriate for IPM training aimed at developing sustainable and ecologically responsive practices. FFS was then developed by FAO, as part of the IPM package, to equip farmers with knowledge and skills for judicious pest control decisions and practices, to enhance their understanding of rice crop ecology (Kenmore, 1996). This innovative ecological approach to training farmers about IPM was based on four principles: growing a healthy crop, observing the field weekly or regularly, conserving natural enemies, and farmers becoming experts. All of these were developed to encourage use of farmer practices (knowledge and skills) integrated with modern scientific rice ecological practices. A more practically oriented approach that focused on farmers' practices in sustainable rice production was necessary. In the case of Indonesia, there was a need to avert the situation to save the entire ecosystem. Rice was (and is) a major cash and food crop for the entire population. The bad effect of pests on crops and pesticides on the environment were pressing issues affecting the entire rural population. The problem was felt, shared and understood in the same way by farmers, scientists, and government, therefore an opportunity lay open to engage collective action by all parties to cope with the problem of indiscriminate pesticide use.

In groups of 25-30, farmers were trained in season-long, field-based, and hands-on sessions. The training followed the phenology of the rice crop in a chronologic way. On a regular (weekly) basis farmers observed interaction of elements in the cropping system and the effects of these interactions on pest population dynamics, hence by implication on the ecosystem. A combination of technologies makes IPM a knowledge intensive field, and introduction of IPM innovations necessitated learning by doing. The principles of adult learning, as described in the theory of andragogy (see Knowles and Associates, 1984; Moss, 1983) and the experiential learning cycle (Kolb, 1984), formed the foundation that inspired the process of social learning by doing and self discovery at the root of FFS. Through engagement with IPM and experience generated during FFS training, farmers became experts in their fields (Kenmore 1991), leading to their development of practices better suited to specific local contexts. FFS was, therefore, the most appropriate way to work on the complexity of biological challenges caused by indiscriminate pesticide use. This was very possible in the case of South-east Asia because practice (as a collective need) preceded the establishment of a curriculum. What we see formalized as FFS in theory developed out of ongoing practice, i.e. practice created theory. Activities were not predicted before hand, but evolved gradually.

FFS training in IPM was practice-based, with rice farmers, scientists and government officials identifying with, and sharing, the prevailing problems of indiscriminative pesticide use, and focusing attention on the same objective of reduced pesticide use for restoration of the ecosystem and sustainable agriculture. The strength of FFS as an appropriate method of training lies in its more practical orientation. This accords well with the emphasis placed by the anthropologist Jean Lave on the role of practice as a basis for learning (Lave, 1995). Modifications of IPM to suit specific farms, taking into account socio-economic, political, cropping and farming system differences makes IPM an evolving technology, which adapts itself to changes in practice (e.g. according to shifts in pattern of pest attack or changes in market demands). Pest control strategies are never final but based on rolling adjustments. Scientists do not have an exact answer, good for all time or places, and there is no text book that tells how to make dynamic, adaptive changes, except in principle. The specific role of FFS is to provide a practical framework through which generative, adaptive and observation-based learning can develop, specific to local problems and opportunities. The earlier approach in extension was to instruct farmers in correct procedure. In an IPM context it becomes clear that farmers need research skills more than authoritative knowledge because what they need to know is generated through engaging cognitive capacities and embodied skills via concrete actions. They need to know how to do something, rather than know that something is the case. Owing to its success in Asia, the FFS model has been applied more widely since the 1990s. With FAO support, it has become an extension methodology of choice in many African countries, included Uganda. There are few African contexts, however, in which the FFS is applied strictly in the context of the original IPM problematic. We need to enquire whether the emphasis on farmer learning and self-empowerment characteristic of IPM (where it is a technological necessity) survives when FFS is applied more broadly.

1.4.3 Farmer field schools in Africa

FFS is increasingly seen as a possible mainstream extension practice in new (non-IPM) contexts. It is now applied in many different fields - including soil fertility, non-rice crops, livestock and human health (Braun *et al.*, 2006) - but with as yet rather limited prior assessment of its appropriateness to these new contexts (Davis, 2006). As will be shown in the case study material, FFS in Africa tends to be a formalized distillation of Asian activity. Borrowing from Bellah's (2005) discussion of ritual as the basis for much collective action and performance, attention will be drawn to aspects of application in Africa that seem closer to the performance of initiation rituals than the practices of scientific experimentation. The procedure of establishing FFS follows a theory of application, but evidence of practice developing from an unfettered assessment of the local context is harder to find. In the Ugandan case study material, institutionalization of FFS as a normative set of practices precedes action-based discovery learning. Inadequacy in understanding the contexts in which FFS works increases the chances of the model being used in a form similar to a cargo cult. A cargo cult reflects the common error of confusing correlation and cause. Pacific islanders saw that colonial invaders built wharves and piers from which they landed the equipment for their new regime. The islanders then built similar structures, but the cargo failed to arrive. In "cargo cult" science every procedure is apparently followed but due to absence of something essential the experiment does not work! "Cargo cult science" has been discussed in the education field (see Hirsch, 2002) but is an equally present danger in agricultural research and extension fields. In FFS practice, ecology-based and observation-based learning are essential things without which the approach is but a lifeless model. Forgetting, altering or ignoring FFS essentials is likely to render FFS a "softer" variant of older authoritative knowledge generation approaches rather than a means to induce skill through self-learning.

In Africa, the cropping system is mixed and most farmers carry out agricultural production mainly by traditional methods, with only a very few using pesticides on a few selected cash crops. Therefore pesticide use was/is not as critical an issue as it was in Asia, which renders the concept of IPM less appropriate in African context. But because FFS is perceived to be a desirable model for training farmers, IPM becomes a principle around which FFS is modified to suit prevailing situations. Pests are therefore taken in the broadest contexts, beyond insect pests. IPM is conceptualized in different ways that suit prevailing situations in specific areas where FFS projects are implemented. In Zanzibar, for instance, IPM was modified into general crop management, and therefore FFS was established as an appropriate model to provide training in farm management to improve upon subsistence agriculture (Bruin and Meermen, 2001). Under FFS projects established at the East African sub-regional level FFS was used to train farmers in integrated production and pest management (Okoth *et al.*, 2002), integrated production and post harvest management (Stathers *et al.*, 2006), soil and water conservation (Odogola *et al.*, 2003) and integrated soil productivity improvement (INMASP, 2002).

Where FFS is scaled up outside the framework of pesticide use as a critical ecological problem the chances of losing its practical orientation towards actual farmer practices seemingly increase. Curriculum formation processes tend to be taken over by scientists pushing their interests (Nederlof, 2006). Practice and observation-based learning, the essential things in FFS, tend to be excluded. Minus the focus on ecology and practice, FFS is less likely to work outside the IPM envelop. Restoring the ecological emphasis implies farmers revisiting and evaluating indigenous practices, forming a view of farmers' knowledge as being a rich resource in appropriate technology development. This clashes with the agenda of some agencies and extension workers, who fear the loss of authority. This implies that the essence of FFS is likely to be lost in situations where processes of technology development and dissemination are undertaken in a more formal and bureaucratic setting. Studies about vegetable FFS in Sudan (Arwa, 2002), for instance, showed that trainers organized lectures, commented, explained and answered questions raised by farmers. FFS in this setting took on a more formal aspect, undermining the expected emphasis on informal practice-based learning. Instead, agricultural professionals used the FFS as a regular "school" to prescribe to farmers the 'best and recommended' practices.

In West Africa, a region in Africa where FFS was first introduced in 1995 (Simpson and Owens, 2002: 406), research scientists used FFS to pass on the right practices in cowpea production to farmers (Nederlof, 2006). Focal, in such settings, is the dissemination of pre-existing technology packages. This reverses the FFS objective from farmer learning back to technology transfer. Mancini (2006) refers to adoption of IPM technologies among cotton farmers in India, which gives the impression of FFS being used in effecting spread of technologies. Technology transfer has a link with adoption where farmers are expected to take up technologies as prescribed. FFS can be an important forum for prescription, as well as a framework for discovery learning. But this underlines the problem - to be examined below - that FFS can rather readily become a medium for conventional methods to maintain an established status quo, as argued by Richards (2006). "Up-scaling" and formalization of FFS risks losing sight of the practice-based learning objective, thereby diluting the appropriateness of the technique in encouraging technological developments based on a critical analysis of existing practices.

1.5 Focusing the study

This section brings out the central objective and research question(s) of the study and clearly shows the strategy undertaken to answer the questions. It starts with an overview of the technographic strategy used. This is then followed by positioning FFS in the wider realities of formal institutions, communities of learners and curriculum in which FFS is actually embedded, and how these realities influence operation of FFS as will be seen in 1.5.2, 1.5.3 and 1.5.4. It is from the influence of the wider realities that research questions were generated. The approach used to collect and analyse data is also described.

1.5.1 Technographic overview and descriptive/analytic strategy

The present study is an attempt to understand how FFS actually works, in the new, extended (and non-IPM) contexts just described, based on case-study material from Uganda. Such a project is an attempt to analyze what Hall would call an "innovation system" (Hall, n.d.; Hall and Dijkman, 2006). Innovation, a social process of integrating new skills, knowledge, techniques and processes involving a range of actors, is subject to a whole range of social and institutional factors patterning people's behavior and interactions. Innovation systems projects like FFS projects reflect the visions of a range of actors. The overall shaping of the entire venture is the product of the interactions, objectives and goals of different organizations, as well as individuals, involved in the process of developing and implementing the idea. The innovation system described below, therefore, is a network of actors cooperating and working together to develop desired agricultural technology intended to improve the socio-economic performance of Ugandan rural farming communities. Understanding social system dynamics and interactions are an important part of innovation system analysis. Local reality is dynamic and new ideas need to have the ability to respond dynamically to changing contexts. The social system and any set of development projects within it are both open systems subject to change as inputs vary. Different people are affected differently due to the different contexts in play. It is important to look at different layers of social reality within which projects like FFS are embedded because they contribute to the changing nature of projects (interventions, operation, effects and outcomes) at all levels.

In FFS, agricultural technology is a key entry point in enhancing development, especially in agriculture sector. Technology is a focal word in the work of the Technology and Agrarian Development (TAD or TAO). TAD or TAO studies innovation systems (and what could be done when the system is not working well!). The TAO approach to technology is "technographic". Technology is defined as human instrumentality (i.e. use of tools, machines or processes in the pursuit of human objectives). This means that technology necessarily embraces more than tools, machines and processes; it is equally the study people (in the present case, of researchers, extension workers, farmers and other developers and users of agro-technology), and how people and tools, machines etc are bound together in a variety of social and institutional relationships (including, most importantly of course, relationships of production and consumption). To understand a technology, or technology transformation process, therefore, it is essential to grasp the system "in the round" - all key interactions and relationships need to be included. This descriptive/analytical task is "technography"; specifically, in this case, this thesis will attempt a technography of an innovation system (FFS in the Ugandan NAADS context).

The Ugandan agro-technological transformation process is based on a realization that to kick start agricultural technological transformation some "outside" help is needed beyond "getting the prices right" (i.e. market reform). One form in which this kick is being delivered is through FFS (rightly or wrongly) to meet prevailing local needs and demands. To do a satisfactory technography of FFS in Uganda we need to show how the system works (what interactions there are between tools [technical inputs], people [users], and institutions [ways of doing things, rules of the game, social norms]). FFS is rooted by background in pest management [IPM], so comes with a certain kind of institutional "baggage" related to the IPM context, but it is also a group learning approach based on the kinds of claims made by Lave and others about "learning through doing" and embodiment of skill (Lave, 1995; Richards, 1993). FFS claims to be able to form new technical knowledge "in situ" through a modified self-help approach (using local resources of time and energy). The attempt adapts itself to a larger local institutional framework that includes the rural society in Uganda, specific conditions in eastern Uganda (specifically Teso), and the NAADS initiative, among others. This thesis seeks to grasp, describe and analyze this FFS-based process of agro-technological adaptation, in order to arrive at some overall judgment about how well the FFS-based "innovation system" is working. To this end we are especially aiming to describe FFS in terms of what Pawson and Tilley (1997; 2006) call a context-mechanism-outcome configuration (specific to a realist evaluation approach). The basis of this approach is to test for the presence of hypothesized mechanisms explaining how project intervention effects are produced, by what means (how), and for whom, under what specific contexts.

FFS in Uganda was not formally institutionalized, and was implemented in ways typical of the NGO approach to projects. It thus involved intense farmer training on promoted technology, advisory inputs, provision of initial material inputs, promotion of gender equity and some contributions in kind (typical of NGO approaches). NGO approaches claim to target the most vulnerable (poor) members of the community. However, Kidd (2004:139) observes that this is a promise not always or even often honoured. The thesis examines the equity performance of FFS in further detail below. Implementation of FFS was also operationally quite typical of NGO projects. The features of such operations are limited duration, small staff numbers, and restricted coverage; i.e. the activity works with small numbers of farmers in a relatively few specific districts villages and farming communities, but in the hope that success will create demand for further implementation.

Apparently, there is no shared problem that cut across different farmers in the different geographical sub-regions of Uganda. Nor are problems addressed often felt as collective problems affecting all communities where the projects are introduced. A diversity of problems among small-scale impoverished farmers is one of the key problems of poverty. It is hard to build up momentum or critical social mass behind any specific problem, except when it is of a catastrophic nature (e.g. a famine or flood), and then little can be done except supply relief and rehabilitation. A key question for FFS, then, is whether it can locate a shared problem that evokes enough general concern and interest for a critical mass of intended beneficiaries to engage in collective action through some kind of action research process. The diverse nature of the mixed farming system in Uganda makes it difficult to attain a situation where farmers have the same feelings about and develop a common understanding of a shared problem. A rich variety of crops, animals and off-farm activities combine into highly specific stakes that vary across individual farmers, villages, sub-counties and districts. The typical poor farmer in Uganda is a risk spreader. This situation provides alternatives that farmers can fall back on in case of a problem with any specific enterprise. Besides socio-economic variations among farmers, difference in climate, soil fertility levels, topography and rainfall underpin an enormous mass of local variation in patterns of agricultural activities. Each variant poses different problems and engages different interests among the farming community, even within a single group or village. In this context the key to farmer participation may not lie in development of technology but in supplying a richer variety of viable on-the-shelf technologies, i.e. a revised variant on ToT might chime better with poor farmer interests than the idea of in-situ problem solving, where there is little agreement on the basic nature of the important problems to be solved. The practice-oriented spirit of FFS might be better fulfilled, in the Ugandan context, by emphasizing the bazaar rather than science (seed fares might make more sense than farmer participatory breeding, for example). Thus in addition to understanding how FFS actually works in the Ugandan context the thesis also aspires to inform commentary on the best ways of adapting FFS initiatives in the Ugandan context.

This brings us to a statement of an underlying objective in conducting this study: viz. *to make a contribution towards improving use of FFS type interventions in re-orienting the agricultural extension system in Uganda.* Information generated from this study is intended as an input into the decision making process about whether (or how) to use FFS more effectively and efficiently, in enhancing agricultural or rural community development. Many participatory approaches exist and are designed to suit specific problem situations to which they aim to make contributions (see Leeuwis, 2000). FFS aims to promote active farmer participation through groups formed around technology. As FFS gains momentum there is a current lack of clarity on (a) how farmers participate in FFS and (b) how the context in Uganda supports observational learning to build farmers' skills in analysis and decision making. There are successes and failures to report, but no one yet knows how to answer the fundamental question "should this be the general approach for extension in Uganda?" - in short should (and could) FFS be up-scaled?

The focus of this work revolves around the participation process in FFS. Despite participation as a concept or aspiration appearing as a notion guiding agriculture projects in PEAP, there is never a complete and adequate connection between policy principles and the way projects like FFS work out in practice. This is in part because participation can be defined and implemented in so many ways. But it is also because participation (or interactions) and learning cannot be detached from each other; they go hand in hand. The two occur in a cyclical process and are on-going: people learn as they participate and participate as they learn. Learning takes place through participation and is a collective process (Lave and Wenger, 1991). This study pursues a social rather than a more familiar psychological perspective on learning because people are generally interdependent by nature. What people do, say, and think is greatly influenced by interactions within the wider social environment. This leads to the notion of "social learning" - i.e. the idea that knowledge and skill formation are fundamental ways in which social projects, involving cooperation and collective representations work. The implication of this perspective is that the social organisation of FFS projects is a key area for technographic scrutiny. The thesis approaches mechanisms of learning through an account of the social processes apparent in FFS groups.

We begin to seek answers to the central research question about operation and effectiveness of FFS in re-orienting agricultural extension by looking beyond FFS projects to the wider context. To understand operation of FFS we need to explore diverse rural realities that have bearing upon the functioning and outcome of FFS. We capture these contextual factors in a Venn diagram showing the intersection of formal institutional context, 'communities of learners' and curriculum as illustrated in Figure 1. The formal institutional context refers to the whole range of organisations engaged in FFS projects. Frameworks, traditions, activities, mandates, objectives and interests vary across these organisations and influence operation of FFS. These organisations include technology sources (research groups) and those that relay technologies to farmers (extension), thereby connecting communities to development.

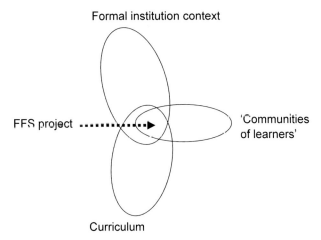

Figure 1: Framework of the study.

Interactions between organisations with related or similar targets are continuous and what happens in one organisation influences what happens in another. 'Communities of learners' refer to villages and districts in which technology is introduced. Social-cultural traditions and the farming practices/systems of these communities influence how farmers respond to a new technology. Curriculum refers to the ways in which information and technology (content) is delivered via FFS to solve farmers' problems. The content influences the choice of teaching method used. Different teaching methods are suitable for different objectives. Although different organisations may have specific teaching methods, different contexts require different teaching methods based on the technology and the type of learners (farmers) in question.

1.5.2 Influence of the formal institutional context on operation of FFS

Institutions guide formulation of policy, which direct people's actions. This does not exclude FFS projects whose key actors; include farmers, extension workers and researchers, have different institutional or organisational mandates to push for. Often, different actors serve according to their interpretation, convenience and understanding of a participatory project, a situation that Mosse (2005) encapsulates as the tendency of different parts of a project system to operate with considerable autonomy from one another. In the context of FFS, where collective action and learning are emphasized, how does each actor then play their roles in ensuring effective operation of FFS? Leeuwis and van den Ban (2004:15) made an observation that actors such as extension workers or change agents working with donor or government organisations tend towards being brokers in the process of mediating or marrying different interests between funders and local people. This necessarily affects how they relate and work with the different actors. Participation emerges from a complex process of negotiation where field workers are subject to competing influences from the organisation they work for, the communities they work with and their own characteristics or needs as individuals. Situations that challenge field workers' desires, preferences and commitments are most likely to make them use their discretionary power to devise strategies to improve their working environments (see Lipsky 1980). This activates a micro-politics of project life based on tensions between and among the aims of different actors.

Contradictions in or competition of interests between individuals and their institutions across all levels ranging from international, national and local creates some pressures responsible for performance being at times contrary to the stated set up of the project. Negotiations and trade-offs are only attainable in an interactive atmosphere that is provided through social learning. Sometimes, however, members of the community who interact with researchers may not provide a true picture of the situation due to personalised or unknown interests. It is a moot point whether researchers realise this, or take this up as a learning point for subsequent projects. Unless explicit attention is given to the intended target audience, for instance, there is a tendency for project staff to interact mainly with already better informed or more powerful and articulate people (Shucksmith, 2000: 215) perhaps because they are easier to access, or because they better realise the advantages of playing the project "game" than the ones in greatest need of the initiative and with the most appropriate knowledge. Interplay between the knowledge, power and interests of various actors contribute to the emergence of contradictions between policy and practice. This is the basic reason for a technographic approach - in relation to FFS, we need to to find out what actually works on the ground.

FFS fits within an integrated agricultural system, and is seen as a process in which different actor interests and struggles are located. It then reflects more of a battlefield of knowledge (see Long and Long, 1992) since it brings people with varying knowledge perspective together. People perceive and make decisions based on the social interactions in which they are involved. Their interactions inform their thinking and interpretation of events. The nature of interaction, therefore, is likely to depend on how knowledge is perceived, negotiated, processed and created among or between actors involved in the process (Röling and Engel, 1990). Institutions and institutional styles depend a great deal on actor interactions and norms of behaviour (North, 1990). Sometimes, institutional style or rules block appropriate use of new approaches. This is why we need to understand the diversity of institutions (Ostrom, 2005) if we are to effect change in research and extension system for development.

Insights that institutions sustain the thinking behind research and extension helps explain why farm related problems are dealt with in specific ways. Research and extension tend to implement projects in the same ways as ingrained in the traditional way of doing things that often replicate or reflect how their organisations function. This is what Douglas (1986) refers to as the institution doing the thinking. Yet, as the saying goes, if you always do what you have always done, you will always get what you have always got. Changing an approach or policy while maintaining the same practices, therefore, becomes a 'new wine in an old bottle' scenario. This is why there is need for institutional change and therefore reformulation of how organisations function to match new demands and challenges (new bottles as well as new wine!). Improving a farming system does not refer to prescribed practices but challenges actors involved to think about interactions, dynamics and alternative models that foster participation for rural or agricultural development. The same applies to FFS if it is truly to transform traditional concepts of extension as a path through which technologies and recommendations reach farmers. FFS serves, therefore, as a model case for the entire question of reformulation of agriculture and rural development agenda in Uganda.

In a situation involving more than one person or organisation, questions of inter-personal or inter-organizational power relations arise. The same is expected in FFS settings, where farmers, extension agents and researchers come together with the objective of collective learning. Power is context bound. The tendency for one set of agents or organization to feel more powerful than another is almost unavoidable. The issue then is not who is more powerful, but how to make use of that power collaboratively to create a more supportive environment guaranteeing the interactions required for constructive learning. How is this handled in FFS? Contexts within which farmers, extension workers researchers and other actors work differ and therefore their knowledge in terms of both content and orientation will differ. Does all this knowledge work better when moulded into a whole, and when each party views the other as constructive partner in development? The idea that knowledge is contextual implies existence of heterogeneous forms of knowledge, but with common perspective to maintain social order. In the process of collective learning, as Keeble and Wilkinson (1999) observe, a base of common or shared knowledge is created and further developed to make up a more productive system. It is a concern in this thesis to examine if, and how, this framework of common or shared knowledge is created under FFS in Uganda.

1.5.3 Influence of communities of learners on the operation of FFS

Learning is (as argued) a social process, although effects are felt and can change at the individual level. (Social) learning is used in two different perspectives: through practice and through facilitation. Whereas practice orientation falls more at the non-discursive end of the participation continuum, facilitation appears more at the discursive end of the continuum. Richards (2006) argues that although the appropriateness of both kinds of participation tends to vary with context, the power of their adequacy lies in the capacity to evoke collective action. He further mentions that a balance between the two would benefit participation around technology. Because FFS is more oriented to skill building around technology, this thesis takes special note of the practice oriented perspective in arguing the relevance of FFS as an approach to promote social learning. Lave (1995), in a paper presented at the 1995 American Educational Association Annual Meeting, pointed out that most ideas of (social) learning are "cultural artifacts" of teaching without a proper explanation (theory) of what teaching really is. Her work shows that learning is fundamental to all forms of participation. Teaching by contrast is a cross-context facilitative effort to make (high quality) resources available for communities of learners. This also implies that teaching is neither necessary nor sufficient to produce learning.

By-and-large other authors (see Goldstein, 1981; Swenson, 1980; Hergenhahn, 1988) have a notion of learning that is close to Lave, but without so clearly distinguishing learning and teaching. Moreover, as various authors in the 'tyranny of participation' debate (See Cooke and Kothari, 2001; Hickey and Mohan, 2004) have pointed out, there is a problematic idea of society behind these models - society seen as a kind of zero-sum power game. The argument goes roughly like this: social learning is most effective (or only effective) in communities where all participants have power equally divided among them. In order to reach that state people have to be taught what equality is. FFS is no exception to this 'bias towards equality' in seeking to reach an ideal state of free deliberation on a level playing field, or to come as close to it as possible - but acts of instruction (often termed 'facilitation') undermine and contradict the notion of equality and strengthen the position of the vocally capable with respect to the vocally less capable. Interactions, and therefore learning based on social practice, are natural processes mainly guided by interests of the learner. However, through facilitation, learning can be directed more to suit interests of researchers and not necessarily towards what farmers want to learn.

Contrasting Lave's practical position with that of many FFS theorizers (see Leeuwis and Pyburn, 2002) helps us to clarify what the analysis of data presented below achieves in technographic terms, as well as providing a basis for suggesting improvement (and up-scaling) of FFS functioning. The separation of learning and teaching clarifies the tendency that FFS encounters to move back towards traditional forms of instruction. Moreover, it provides an opening to better understand the (social) functioning of facilitation in regard to technical issues. Introduction of technology for improved farming practices generally assumes a situation of teaching, which implies making resources available for learners. Yet not thinking through properly what 'communities of learners' (in this case the farming communities themselves) are to be reached and what teaching methods would fit best to that objective may result in increased exchange value of learning independent of its use. The way forward then would be to start with a proper (technographic) understanding of the 'communities of learners'. Such analysis (attempted below) reveals patterns of social interaction (learning) that become a basis for identification of appropriate resources (of an agricultural kind) and a selection of (existing) teaching methods suited to that learning context.

Where there is social interaction there is learning. Participation discourses emphasise and often result in learning. Learning processes, commonly viewed in the context of social learning, are referred to even more often and more explicitly than decision making models (Leeuwis, 2000: 936). In FFS, facilitation is more emphasized than practice/doing, yet participation is more a practice than a discourse. Members of local communities interact with each other and with other stakeholders, in this case researchers/scientists and development agents, to come up with feasible solutions to existing problems; the complexity of farming related problems cannot be effectively handled single handed. However, this scenario works best if each side has a better understanding of what the other can offer, and this understanding can only be realized upon interacting. All actors or stakeholders are interdependent; every one has something to offer from their diverse experiences and knowledge (resources). Experience, theory and practice continuously inform each other in a cyclical process, making social learning a process of praxis. How does this happen in the FFS setting? Some type of information, as Isaksen (2001) observes, can only be revealed and known through interaction, because of its informal and tacit nature and difficulty of transfer through formal channels. How supportive is the learning climate in FFS in promoting effective exploitation of knowledge and intentions from all actors involved, including informal and tacit knowledge? This is not only a question to farmers (often referred to as learners, even in FFS), but to extension workers and researchers, since they too are supposed to be learners, where the FFS model is taken seriously.

Collective learning is not only limited to knowledge and skills but involves generation of confidence, resources, insights, perspectives and procedures on which action can be based. How does this result in action geared towards solving farming problems faced in the communities? Interacting and understanding each others' 'language', situations and setting research agenda is a process that will involve power relations, false expectations and limited capacities (Sutherland et al., 2001:79) for desired development. However, all processes and outcomes of such interactions, irrespective of their negative or positive effects on development, are part of the learning process. Learning is synonymous with change in this context; therefore any change in ways of interacting implies presence of learning. Ineffective communication and system use, attributable either to lack of information flow or to transmission of information that is somehow misleading, distorted or even contradictory, is part of learning also. What we want to know is how this kind of learning (from failures and mistakes) contributes to adjustments in FFS to make the teaching appropriate for all learners?

For Chambers (1997: 97), factors which mislead combine where power is concentrated. He further explains that myth and error are generated and sustained by 'uppers'' impediments of dominance, distance and denial and by blaming 'lowers', and by the strategies of 'lowers' for selective presentation, diplomacy and deceit (p. 88). This affects group outcomes, therefore participatory process discussions, given that highly placed people tend to exert some influence on the form and content of discussions, which in many cases is not fully observed, in such a way that the views of ordinary participants becomes aligned with the views of discussion leaders (cf. Murphy 1990]). So participation creates knowledge states that align with social relations and power balances, rather than being truly emancipatory. Without formally applying a realist Context-Mechanism-Outcome framework (cf. Pawson and Tilley 1997) recent work by Humphreys et al. (2006) suggests that the mechanism is to be found in the pre-adaptation of the social order to leadership effects. Participant consultations in the African context they describe (Sao Thome) adapt themselves, in a patrimonial culture of respect for leadership, to leaders' preferences, irrespective of demographic characteristics (gender and age). In other words, participants align themselves with the wishes of discussion leaders, even where these leaders are carefully randomised to include representatives of commoners as well as elites, women as well as men, and youth as well as elders. In the short term, the familiar social order survives and imprints itself on participatory discussion, even when new leaders emerge. If these results prove robust then FFS faces a more severe challenge to change socially-embedded ways of thinking than often appreciated. The good intentions of the designers of such programmes will count for little if the prevailing institutional culture assigning people to categories "rich" and "poor" remains unshaken.

1.5.4 Influence of curriculum on operation of FFS

Curriculum connects the communities of learners (farmers) with the formal institutions (organisations) that generate and deliver information and technology (content). Technology is the central factor connecting three realities: organisations, farmers and curriculum. The nature of available content dictates how best to deliver it to users in a way that makes it more useful. In FFS, participation is implied in the curriculum. Use of FFS is one way to aim at increased participation in society through involvement in group activities and enhancement of farming capacity (via technology). I focus on society and technology because the two are interdependent. Interactions around a given technology influence its availability and use. Improved farming practices, productivity, and ultimately poverty, are often viewed through use of new and improved agricultural technologies in form of crop varieties, planting methods, etc.

Agriculture is a way of life to many subsistence farmers (Richards, 1993) and farmers are in constant search of ways in which to improve upon their lives. Technologies offer opportunities for improved agricultural production and hence improved life. However, efficient use of technology needs some element of training (i.e. teaching, see above) in how best it is used, especially if it is new and external. This is where technical competences of farmers need to be developed to use technology. Interests and mandates of organisations influence strategy or method used in teaching farmers about technology. There are many teaching methods, but each of them is suitable in specific contexts and in meeting specific objectives (Braakman and Edwards, 2002): knowledge generation, change in attitude, building skill or change in behaviour. In FFS, action oriented teaching methods (i.e. practical teaching) are important because the aim is to build farmers' skills in improved farming practices. Facilitation is often taken as the most appropriate teaching method (see van de Fliert, 2000) to encourage action-oriented learning by self-discovery. The objective is generally achieved through using a combination of methods depending on the situation: complexity of the content, knowledge level and needs of the learners, and competence of trainer, among others. Teaching method used influences farmers' uptake and integration of new or improved technology into their real social and farming context. This is why it is important to build from what learners already know (see Posner and Rudnitsky, 2001), which then implies flexibility in choice of the teaching method that connects the learners' real context to the content. Curricularists (Tanner and Tanner, 1995) advocate that the curriculum be adapted to societal needs but warn against taking it as a ready-made process.

By paying attention to farmer participation in the case study material below we look at participation from the perspective of skill building for communities of learners representing the grassroots rural farmers. Such orientation towards local realities provides a basis for holding development interventions such as FFS accountable for their relevance in the local contexts in which they are introduced. Particular attention is paid to local processes at work in curriculum formation, and whether the resulting FFS curriculum is device more for the convenience of teachers rather than learners, a possible criticism with educational implications wider than FFS alone.

1.5.5 Specific research questions

In order to find out how institutional context, communities of learners and curriculum used influence operation and outcome of FFS in Uganda, two things seem especially important. The first is the need for a better understanding of the processes through which policy is percolated to reach the field. The second is to attain a better understanding of realities in the field where FFS projects have intervened. Development through FFS cannot be read from policy documents and is not a straightforward application of some principles of participation. What is required is meticulous analysis of specific activities and processes. This leads us to spell out a set of more specific questions guiding the present study:

a. *How do actors shape operation of FFS?* This question aims to identify the actors (organizations) involved in the FFS operation, the roles each plays, and how institutional contexts influence the way FFS has been implemented and operated in the field in Uganda. In this way we will hope to explore the position of farmers in the larger participation process.

b. *How does FFS as a problem-solving process fit within the existing specific local contexts of the communities of learners (farmers)?* This question is directed at the contextualization of FFS. It requires some discussion of the existing social and farming practices of the farmers, and how FFS interventions influence farming and the social system within which the targeted farming communities are embedded.

c. *In what ways does the curriculum boost participation to build farmers' skills in better crop management practices?* This question seeks to throw light upon the different methods used in teaching farmers about new technologies, and how these methods engage with and are suited to prevailing farmers' conditions.

1.6 Approach to the study: unravelling the operation of FFS

The technographic approach is used to understand and analyse actors, activities, processes, and the interactions between technologies or interventions promoted in FFS and among communities of learners. Technography requires a meticulous design and specific methods for comprehensive data collection and analysis. The section that follows gives an account of the design and data collection methods used.

1.6.1 Research design and data collection methods

To find out how FFS re-formulates the agricultural extension system in Uganda and links policy and practice necessitated a case study (see Creswell, 2003; Yin, 2003) as the most appropriate strategy. We need to know the 'what', 'who', 'how' and 'why' in tracking participation processes, and hence the thesis relies upon a research strategy providing detailed analysis of a full range of activities/aspects and procedures used in FFS. Digging towards the underlying rationale leading to observed outcomes in FFS required time. For a period of about 3 years (January 2003 to end of 2005), the study was conducted through an intense and prolonged contact in the field (actors and context of FFS). The long duration aimed to capture life situations in order to explicate ways in which actors in particular settings perceive, understand, account, take action and manage day to day situations (see Miles and Huberman, 1994), therefore how they interact with each other.

The study began with an exploratory phase providing an overview of distribution and use of FFS across Uganda. This was followed by a detailed investigation phase employing some elements of an ethno-methodological approach (see Babbie, 2001). A central focus on the rationale underlying the patterns of interactions between actors in the FFS provided appropriate ways to answer certain research questions. Residing with the community and sometimes working with farmers in their fields was an important means to establish research rapport. Through these interactions, I was able to identify and establish links with key informants for my study. Key informants are often considered critical to the success of a case study (Yin, 2003: 90): they not only provide insights into the matters in hand but also suggest sources of corroboratory evidence and serve as a sounding board for explanatory hypotheses that make sense in local understanding. Discoveries based on participant observation and interviews were then viewed analytically to guide the next move in the field, as suggested by Miles and Huberman (1994).

Using different methods, data was collected as an outsider before gaining confidence of my respondents, and then as an insider, after establishing adequate rapport. As an outsider, I mainly used group interviews, observations, discussions and photography. As an insider[3] I engaged actively in various project activities and interacted more intensively with respondents. During the insider process individual interviews, participant observation, conversations and focus group discussions were mainly used. Direct observation and participant observations are two different approaches. While (direct) observation is passive, participant observation is active because the researcher takes part in some of the activities or cases being studied (Bernard 1995; Dewalt et al., 1998). Participant ranking was also used to identify local priorities.

Use of an audio voice recorder, especially during the initial group and individual discussions, allowed collection of much material in a short time. Unstructured and open ended questions guided the interviews, discussions and conversations. Silverman (2001: 87) mentions that unstructured and open ended questions provide authentic data giving insights into people's experiences and constructions (i.e. artifacts of questionnaire construction are avoided). Although groups often preferred their group leaders to respond on their behalf, points of agreement and disagreement were easily noticed and created avenues for follow-up questions at individual level later. Observing expressions used as people responded to my questions created chances for follow up and posing of more questions.

Photography (by informants) was used as a strategy to collect data, in the form of stories around events that happened in my absence. Each member took a picture of two objects of their choice. The photographs were intended to provide a point of discussion: the issue was not the photos but the events and stories that lay behind them. However, this was expensive,

---

[3] My involvement in various project activities and meetings enabled me to access some information that seemed confidential. However, this was after project implementers accepted me as part of them in aiming to improve upon the projects. Through the same meetings I was able to offer feedback to them about what I thought could be done to make improvements in the operation of FFS projects.

and did not work well, even if it served to create stronger relationship with some FFS groups and individual farmers. But it did provide some important insights into how FFS actually works. Farmers and facilitators were not familiar with the technique, and did not take it seriously. They used it instead as an opportunity to take pictures of family members (perhaps an important indicator of what they found truly significant). An equally important finding is that some facilitators dictated what farmers should choose to photograph, and even used the camera more often for their own purposes instead of project business. Pictures ended up being choice of the facilitator, probably the reason why so few farmers could give a story to back up their choice. Here is a stark indication that despite the basic orientation of FFS to popular empowerment, facilitators assume they have the right to direct the gaze of farmers towards objects of their own concern. The research strategy might be deemed more revealing in what it failed to do than if it had succeeded. After a discussion around what had been photographed each farmer was left with his/her pictures.

Frequent visits were made to farmers' fields and homes, initially on appointment and later without appointment. Invitations and sharing a drink with some farmers, especially in some evenings, created a relaxed atmosphere. Appointments tended to create artificial and unsustainable behaviour by respondents. Helping respondents in their fields while collecting data strengthened dialogue and informality between farmers and researcher. It was challenging, however, in the sense that the fields were far and farmers returned home late very tired. There were also moments when the researcher was asked questions by some farmers and had to carry out on-the-spot extension and advisory work in the fields.

Formal meetings, especially planning and evaluation workshops, provided supplementary information. The researcher took part in most activities and sessions either as a participant, facilitator or resource person, and interacted with farming community, FFS farmers, facilitators, local leaders and project bosses formally and informally. This was possible mainly due to rapport established between the researcher and the project personnel, as well as with farmers and facilitators. Sharing ideas and information with the project personnel about farmers' perceptions and expectations of the project made the researcher more readily accepted by project staff. Because of her suggestions in some cases, she was invited on several occasions to take part in various project meetings: planning, review, training and evaluation meetings.

For data Analysis, instead of developing chapters by project or district, it was thought more worthwhile to use a common theme approach. This not only minimises repetition but also gives a clearer overview of processes undertaken in FFS. Transcribed data from the field were reduced and sorted into emerging themes and patterns to describe and explain processes and activities in FFS, using the context-mechanism-outcomes framework of interactions in FFS. To convey a flavour of the feelings of the different respondents, some excerpts based on their own wording have been built into the analysis as direct quotes. Use of direct quotes from informants is recognized in qualitative data analysis. However, Bernard (1995: 363) advices qualitative researchers to avoid lengthy quotes that lack analytic value. He makes the point that data do not speak for themselves, and the researcher has to develop his/her own ideas and analysis about what is on-going, using (as here) quotation as a means of illustration.

1.6.2 Selection of cases and study sites

Five FFS projects were purposively selected as the cases for this study. In the selection process priority was given to (1) engagement in different topics, (2) accessibility of project districts, free of interruptions due to civil wars/rebel activities, (3) readiness of project implementers to work with the researcher as partners and learners. Stage of implementation was also a factor considered in compiling information to give a broader picture of the FFS cycle: some projects were ending or had just ended while some were midway and others were at initial stage. These projects were:

a. Integrated Production and Pest Management project/scheme (IPPM) implemented by FAO
b. Integrated Pest Management (IPM) project implemented by Makerere University (MAK)
c. Integrated Soil Productivity Improvement (ISPI) under Africa 2000 network (A2N)
d. Integrated Production and Post-harvest Handling Management (IPPHM) implemented by CIP
e. Safe Pesticide Use and Handling (SPUH) implemented by MAK

Study sites
Selection of cases had implications for study sites. While some projects/cases were confined to Uganda, others operated beyond Uganda. Some even shared districts of operation within and outside Uganda. IPPM-FAO and IPPHM-CIP were regional projects that operated within Uganda, Tanzania and Kenya. Although the IPPHM operated in only Soroti district in Uganda and IPPM-FAO covered Busia and Soroti, both cases/projects operate within the same regions or districts in Western Kenya (Busia, Bungoma and Kakamega) and Northern Tanzania (Bukoba). The other projects operated only in Uganda. ISPI-A2N covered Busia and Tororo districts. IPM-MAK covered Kumi and Iganga districts, while SPHU-MAK, whose initial activities began unfolding in 2004 when I was still undertaking fieldwork, covered three districts of Mukono, Mbarara and Kiboga.

Soroti and Busia were the principle study districts because they were the pioneer districts in using FFS in Uganda and each shared two cases/projects. This was followed by Tororo district. Data collection was not limited, however, to the selected cases and districts. But frequency of interaction with other districts was lower because of their lower levels of FFS activities. A combination of districts with various FFS projects at various stages (beginning, on-going and ended) provides a broader spectrum of activities, processes, interactions and learning in FFS. The study site was Eastern Uganda, because of the location of the three principle study districts of Soroti, Tororo and Busia (see Figure 2).

Historical background of Eastern Uganda - the principal study site
Eastern Uganda is constituted from Teso, Bugisu and Busoga regions. These regions are occupied with the largest and main ethnic groups: the Ateso, Bagisu and Busoga. Teso region, the principal site of this study, was originally one district (called Teso district) and divided into

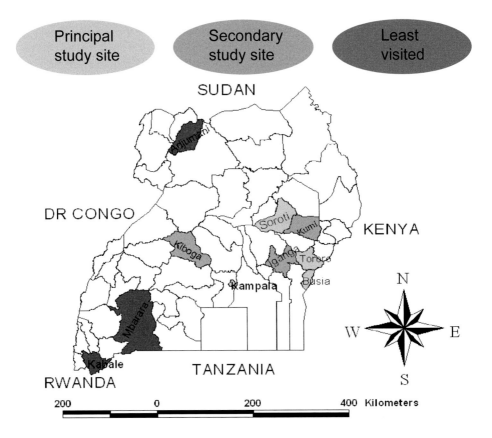

Figure 2: Map of Uganda showing study sites by districts.

two: the north and south. In 1980, north Teso became Soroti district and south Teso became Kumi district. The two districts (Soroti and Kumi) have been further split into more districts (Rwabwoogo, 2002). North Teso presently constitutes the districts of Soroti, Kaberamaido, and Katakwi while the south constitutes Kumi, Pallisa, Busia, and Tororo districts. North Teso is more agricultural in orientation while south Teso is more oriented to trade probably because of its closer proximity to the Uganda-Kenya boarder, especially Tororo and Busia districts. The soils in the north are less infertile than those in the south though the whole Teso is characterized with sandy soils. The landscape of eastern Uganda, particularly Tesoland, is generally a low flat plateau divided by swampy valleys and covered with savanna grassland and patches of deciduous woodland and thicket. Production is based on annual crops in an environment characterized by light and infertile soils, heavy precipitation in two rainy seasons and a fairly prolonged dry season that stretches from December to March.

The Teso farming system was based on ox-drawn power and a basic cropping system of finger millet and cotton under a fairly high population and livestock density. However, the system has undergone a process of long term change resulting in insufficient food production, persistent food insecurity and poverty. A study conducted by NARO/DFID (Akwang *et al.*, 1998) to assess the agricultural research needs of the region revealed that massive de-stocking as a result of cattle rustling, prevalence of HIV/AIDS, insurgency and insecurity, unpredictable and unreliable climate, and climatic change, greatly contributed to inadequate food production in Teso. Cotton became the major cash crop in Eastern Uganda after its introduction during the colonial era but production latterly reduced due to rising costs of production and breakdown of market structures, among other factors. Millet, sorghum, cassava, cowpeas, maize, groundnuts and sweet potatoes - the other major food crops - are increasingly cultivated for cash. Mangoes and oranges are also abundant in Teso, though not major sources of income mainly due to absence of market infrastructure.

Teso has suffered a great deal from civil strife, greatly affecting the socio-economic welfare of people in eastern Uganda (WICCE, 2002; Henriques, 2002). Teso was once a very productive cattle region, but rustling during the 1980s and early 1990 led to the general deterioration of farming systems in a once very productive area (Isubikalu, 1998: 92). Agricultural production was seriously curtailed by absence of the cattle that previously provided draught power for land opening, preparation, planting and sometimes weeding. Unpredictable weather with longer periods of drought, rebel activities and lack of farm inputs minimized agricultural activities, thereby creating more frequent food insecurity and recurrent poverty. Especially in the rural areas, people needed all kinds of support to help them survive: the situation is gradually becoming better. This is probably why a good number of NGOs are in Teso (especially in north Teso). Because the economic activity is mainly agricultural production, most effort to improve rural life focuses on transformation of agriculture as stipulated in PMA/PEAP.

1.7 Organization of this thesis

This study is built around a technography of FFS in Uganda (cf. Archibald and Richards, 2002) and assumes a realist approach to evaluation (Pawson and Tilley, 1997) approach to give an overall analysis of an FFS-based innovation system at work. FFS relates technology and society through participation. Participation is the main element through which the conventional agricultural extension system is being re-oriented in Uganda to reduce poverty. Learning, as a social process, is fundamental in all forms of participation. An account of FFS is given up to the point it was introduced in Uganda. FFS as a mini-program to re-orient agricultural extension uses technology to enhance participation in society. Operation of FFS in Uganda takes a different direction from the Asian model and is influenced by the realities in which FFS is embedded. The realities are broadly categorized into three: mandates and frameworks of organizations engaged in FFS implementation, the existing social and farming practices of the communities of learners, and the way the curriculum is used. Understanding the local context of the farming communities in which projects are to be introduced is very important, yet often ignored. Such analysis provides a basis for a discussion about the organizational functioning of FFS interventions, and the way such interventions might be improved.

The present chapter has served to introduce the main issues. Chapter two looks at institutional elements on the donor-researcher scene and emphasizes the need for organizations to reformulate or change their way of functioning if innovation-based agricultural extension programs like FFS are to enhance participation for development. It is shown that in FFS operations, actors maintain prior organizational approaches despite current community challenges, priorities and demands being very different and dynamic. Findings reveal that prior institutional mandates and the interests of participating organizations strongly shaped FFS implementation in the field. The same mindsets were maintained across time and space in spite of a difference in context. It is thus a major challenge for FFS reform to engineer organizational contexts in which mindsets can more readily change to meet the current demands and realities of Ugandan farming communities.

Chapter three brings out the centrality of technology in guiding activities on the ground. It explores, through examining the technologies covered by FFS in Uganda, how farmers and facilitators are mobilized around technology. Project implementation at local level focuses more on elite farmers mainly because most technologies are more suitable for commercially oriented farmers. Mobilization of elite farmers leads to exclusion of the majority, which undermines development projects intended to benefit marginalized rural farmers. Emphasis on technical content during training did not prepare facilitators to adequately connect with the farmers' prevailing situation. For greater relevance of FFS projects and increased participation of farmers there is need to re-think choice of technologies to be introduced, facilitator training (process and method), and the mobilization method to make them suitable.

Chapter four analyses the performance of FFS in linking with the capacities of the communities of learners. It discusses the social and farming practices in the case-study communities and how these realities relate to project interventions. The variation in farmers' response to technologies introduced reveals the importance of understanding and analysing what happens in client communities, and offers insights into more appropriate alternatives to promote participation. The chapter shows that in spite of a desire to learn, farmers often fail to take up technologies due to technical and social factors in which they (farmers) are embedded.

Chapter five is related to chapter two, but with emphasis on the local organisation of FFS on the ground and the link of the organisational features on the ground to other activities. It describes the internal organisation of FFS, integration with higher structures and linkage to other local activities.

Chapter six explores the central idea that investment in local infrastructure may be a prior requirement for successful application of agricultural science. In other words, scientific infrastructure needs alignment with existing local infrastructure. If FFS is to play the role of teaching farmers appropriately, agricultural research and extension need a properly designed and organised channel to reach farmers. Change in frameworks of organisations may not need only training but new organisational and material structures on the ground as well. As those who have studied post-war reconstruction have discovered, physical repair of damaged bodies may be a pre-requisite for settling the mind. Neither science nor trauma counselling can serve as substitutes for missing material requirements, such as health care and roads. FFS cannot cause development - the thesis concludes. It can only play an effective part as an element in a well-designed total strategy for the empowerment of the poor.

CHAPTER TWO

Development institutions and Ugandan FFS: a technographic sketch

2.1 Introduction

The Farmer Field School approach is a way of working with farmers on learning activities to produce *in situ* useful knowledge, developed first in the context of IPM, but now applied by FAO and others to a wide range of agro-technical problems. It is therefore a process and outcome of interactions involving a range of actors. These actors are governed by a range of institutional ties (values, norms, interests, traditions) patterning styles of interaction (Douglas 1986). Institutional ties, therefore, play a key role in influencing what is observed as FFS in the context of Uganda, where research and extension have long had an upper hand in deciding what is "best" for farmers. Uganda has increasingly adopted the FFS model on a large scale. Since its first introduction in 1996 by FAO-IPPM project on cotton in Eastern Uganda Soroti, FFS has been taken up and used by different projects and organisations to cover different crops and topics (disease, pest, post harvest handling, pesticide handling and soil fertility) across the Northern, Western and Central regions of the country (see inventory in Annex 1). The strength and widespread use of FFS as a promising agricultural extension model, as explained in chapter one, is vested in the participatory and location specific nature of operation.

It is important to look at the FFS approach/model in action to get evidence of what practically happens on the ground, identify areas of improvement and devise feasible ways to accomplish reforms of the system. However, we should make clear that this thesis is not a formal evaluation. It is a technography, in which the first aim is descriptive accuracy using a valid observational methodology. A second aim is to undertake analysis of the technographic data sets and to draw evidence-driven conclusions. In relation to the technographic strategy for this thesis, in which there is a central focus on interaction of people and technology in FFS, this chapter describes institutional elements/actors in Ugandan FFS, and traces the external/internal adaptation process required to set up FFS in Uganda. The chapter identifies and brings out the role of various actors at international, national and local levels, and goes on to explain what actors do and how they relate/interact with each other. It also analyzes the roles of actors and considers how their interests and objectives shape FFS. It is generally shown how momentum builds up to shape FFS.

The overall point emphasized is that while FFS came with a participatory background (from IPM) this does not translate well to either the non-IPM or Ugandan context, and the net result is a system that perhaps repeats some of the top-down mandatory and instructional failings of earlier agricultural extension systems like T&V. Traditional structures and functioning of research and extension remain strong in and shaping of interactions and outcomes in FFS, as will be seen.

2.2 Ugandan FFS and the role of international institutions: a coup?

Projects and priority setting are mainly set at macro level in response to areas where donors and governments put emphasis. On this basis, FFS donors promoted and supported farmer training in crop protection. Problems faced by rural or farming communities were identified by actors at international level without consulting the farmers for whom development was to be delivered. Emphasis on the crop protection theme was based on a perception of donors and researchers (actors at international level) concerning agricultural and rural development. Through proposals and negotiations between donors and scientists at international research institutes, feasible solutions towards managing perceived problems were designed. Decisions made from above were often rigid and provided no opportunity for input from the lower level actors for whom development was meant. Much as interests from grass-roots may not have had a great effect on the research and development agenda at a macro level (Byerlee, 2000), priorities set at higher levels often assume a homogenous situation that consequently tends to ignore interests or realities on the ground. Communities are heterogeneous in what they do, how and what they regard as a priority or of interest, so treating them as homogenous risks projects being less relevant in some cases. Fitting the local context within project priorities is one way of improving upon the relevance of FFS.

Projects like FFS provided platforms in which mandates of institutions with similar or related activities at macro levels were negotiated. Funding agencies set the pace and direction (through conditionality) of FFS projects. International research institutions (CIP and FAO) with similar objectives then pursued the projects with donors. In this process, researchers at the international level reformulated the original objectives of the donor. The researchers - as will be shown - often reshaped a project to suit existing objectives and interests in their institutions. As this process took shape, researchers at CIP and FAO then linked up with other institutions (research and extension) at national and local level for implementation.

It is important to care about realities of farmers if agro-technology "beyond the market" is to contribute to poverty alleviation. Caring about farmer realities means knowing farmers, and hence embarking on active interaction with them in designing projects meant for development. This was why enthusiasm for participatory approaches arose. But recent research has suggested that even in the context of participatory approaches, farmers' interests are manipulated and instrumentalized to legitimize already set objectives and decisions of projects. In some accounts (Mosse, 2005) farmer participation is seen to reflect power relations among farmers as well as strategic adjustment of stated "needs" to match project deliverables. In projects with already set objectives farmers are made to feel that the interventions (or technologies) on offer through such projects are a requirement to gain support for solving prevailing problems. It is as if assenting to project assumptions is a kind of text farmers need to pass to gain access to a further development aid benefit stream. "Success" in aligning "needs" with deliverables then continues to maintain the way the development organization operates. Participation in this sense (so it is argued) is essentially no different from wells, roads or dams. The tail wags the dog.

The starting point of this chapter is a prediction by Richards (1990) - that participation (which he terms populist development) comes in two forms, supply-driven and demand-driven, and that the supply-driven form is likely to dominate the demand-driven form, due to political imbalances (a deficit of democratic accountability in poor countries). Mosse (2005) seems to confirm this prediction. Our focus here is to ask is it simply that "money talks", or are there other factors negating a demand-driven approach? The chapter explores the extent to which it is already built in to the logic of development institutions that "top down" thinking takes over. In short, we shall consider evidence that it is the "institution doing the thinking" (Douglas, 1986), and that what is needed is a different (non-hierarchical) institutional design within which to deliver participatory initiatives such as FFS.

First we shall indicate how "institutional thinking" works. Farmers implement FFS project activities yet project designs do not take into consideration farmer participation in problem diagnosis, priority setting and choice of technology. With minimal care about appropriateness or relevance of projects to local contexts, activities or roles performed by all actors risk overwhelming the FFS implementation process. What we observed was that each actor played his/her role as guided by mandates, interests and objectives of "home" institutions, with overall priority being accorded to the "top dogs" (i.e. to the objectives of international researchers). In promoting institutional interests, actors made farmers believe that the problem being addressed and the way of dealing with it, as selected by the researcher, was very important. This tendency kept such institutions looking at all contexts in the same way, and consequently ensured they were tied to just one way of working with farmers.

Following the cycle of FFS projects from conception to implementation, it was clearly observed that actors at higher levels ignored realities on the ground, a typical feature of the top-down implementation FFS was meant to replace. Priority was given to interests of CIP and FAO researchers as technology sources. This revealed repeated disconnection between the principles and practices of participation for empowerment that FFS claims to promote. He who pays the piper calls for the tune. Eventually, researchers care more about pleasing their donors (and themselves) and not necessarily the farmers who were targeted. Instead, farmers were "schooled" to fit in with researchers' programmes. Often, farmers did not know the funding agency and could neither negotiate for their interests nor influence the projects to suit their prevailing local context. They had little or no sense of rights of ownership, but knew the international institutes (FAO and CIP) as the source of technologies and funds, as preached by local and national actors. They begged rather than demanded.

Gearing actions of organizations at national level towards relaying technologies to farmers without taking a step back to reflect on appropriate organizational functioning at all levels of the system reflected the usual conventional institutional functioning. This is well captured in the statement that if you do what you have always done you will always get what you have always got. Doing things the same way kept institutions thinking the same way. Change in thinking and change in functioning have to go hand-in-hand if the desired interactions are to be realized in participatory programs like FFS.

Absence of farmers' voice in identification and prioritization of problems at higher levels provokes many questions. One fundamental question is how international donors and researchers choose appropriate technology packages and delivery mechanism without analyzing farmers' realities. To what extent do they expect participation in the field when functioning with the same values as before? Whose interests are the priority in FFS projects? Is it the interests of the farmer, researcher or donor? In spite of absence of farmers' voice in articulating problems, technologies and methodology farmers take up anything that comes their way as an opportunity with hopes of making a difference in their lives.

The argument in this chapter is that the way in which a new technology or idea fits a local context and how it is introduced influences engagement and acceptance from the targeted community. Much as researchers ignore the realities on the ground, they cannot run away from them. Instead of taking the initiative to analyze problems with targeted communities in order to help make them make the right choices they find it easier to make assumptions about applicability of technologies in local context. These assumptions are built in to what they do, know and want to do - they are assumed in order to ease what the agency wants to do. How do we reverse this institutional coup, and restore democracy to FFS?

A major change in the planning and budgeting approach espoused by most agencies is required. The most appropriate way to understand or analyze a community is through working with it, i.e. by actively engaging in a local process of problem review. Choice of areas with special difficulties as operation sites might be one strategy to position and fit projects to local conditions. This is in effect to seek the local equivalent of the "IPM problem". What problem is locally so pressing that there is broad agreement about the need for a solution? The wider history of FFS suggest that only then will there be real drive and local commitment to engage in the hard work of acquiring new and unprecedented technological skills and solutions. Success in finding "key problems" (and not "feasible solutions") would also imply that any technological solutions would be taken up by many. In short, if the development is meant for local people, why cannot development organizations develop procedures to match their objectives with farmers' needs from below rather than farmers matching their needs with development project objectives as handed down from on high? Consonant with a shift towards greater democracy, FFS, as applied in Uganda, needs to be reorganised around problem-seeking methodologies as a prelude to the implementation of solution-providing methodologies.

2.2.1 The role of funding agencies

FFS methodology is being aggressively promoted by donors because of success stories from Asia (Davis, 2006). The aggression is reflected in the way FFS is being advocated for almost all contexts, despite the evident fact that situations, interactions and needs differ. Emphasis around crop protection already inclines projects towards pest management technologies whose most appropriate, if not sole, delivery model is known to be FFS, since it worked so well in Asia. Having a requirement for pest management and FFS then serve as a condition to be met by institutions that seek financial support from such donors. The International Fund for Agricultural Development (IFAD) supports, on one hand, use of biological control, host pest resistance and tolerance technologies. On the other hand, the Crop Protection Research Program of the UK Department for International Development (CPR-DFID) emphasizes reduction of pest impact in herbaceous crops in forest agriculture. Due to the conditions attached to the financial support, only institutions with interests fitting with donors' interests could apply, predetermining technologies and delivery or training model at donor level.

A focus on the pest management theme as emphasized by FFS funding agencies has a direct implication that pests be seen as major constraints to agricultural production and therefore that Integrated Pest Management (IPM) will be the most appropriate strategy in pest management. As we saw in Chapter one, FFS was developed out of a need to mitigate the problem of indiscriminate pesticide use that developed as a result of inadequate extension approaches in training farmers about IPM. Advocating for crop protection through IPM automatically brings in FFS into the picture, owing to it being perceived as the most appropriate model to train farmers in IPM. According to this arrangement, the IPM Global facility, hence FAO, becomes the most suitable candidate to steer the design and implementation procedures, because of experience of FAO staff with IPM-FFS. Partner selection for FFS projects at international level becomes clearer. Competence gained by FAO in the design of FFS implementation procedures assumes that the organization understands and takes into account the realities of the local context where the projects is to be implemented. Because FAO is used to FFS, and wishes to see it spread, technology transfer (ToT) based on success elsewhere has (ironically) become the basic model for the spread of what was once seen as a "participatory" alternative to ToT. As participation is up-scaled, with international donor funds and enthusiasm, it risks turning back into the very model of innovation diffusion it was meant to replace!

Let us look at how this has worked in the Ugandan case. CIP pro-actively carried out lengthy negotiations with CPR-DFID (the funding agency) to show cause for the need to have crop protection projects on sweet potato - a deal which led to a call for proposals on similar lines. One member on the implementation team at the international level revealed that lengthy negotiations to convince CPR-DFID about the importance of sweet potato IPPM, operational from 2002, started in 1998[4]. In the broad scope of reduced pest impact on herbaceous crops (a CPR-DFID focus), CIP's interest was automatically catered for, since it lobbied and pushed for the project from the very beginning. The IPM global facility is also pretty much an automatic partner in IPPM, since IPM is implied in IPPM. In this case, shaping of FFS projects began from CIP, which shows evidence of the way that scientists/researchers are at the forefront of lobbying for projects that fit with what they are already doing or plan to do. Between the funding agency and the scientists within CIP or FAO, who keeps the others on their 'toes' to ensure that interests from the lower levels are catered for? This created room for a hypothesis to be formulated: in the implementation of FFS projects, funding agencies and researchers at international level care more for each others' interests than they do for farmers' realities, priorities or interests.

---

[4] This information was collected during an informal meeting with one of the project implementers at the international level during an evening in one workshop.

The Rockefeller Foundation, Carnegie Corporation of New York, and World Bank consortium that funded Innovations at Makerere for the Community[5], commonly referred to as i@mak.com, was biased towards effective contribution of selected African universities and governments to social, economic and political progress (MISR, 2000). The goal of i@mak.com was to develop Makerere University as an institution of excellence in higher education in the area of decentralization through training, research & development and dissemination of research findings appropriate for community development. With the three funding agencies of FFS projects in Uganda, we see how negotiations between actors at international level contributed to FFS projects at the lower levels. The next sections single out the actors carrying on the projects from the international level down towards the lower levels, and how they did it.

2.2.2 The International Potato Center (CIP)

CIP is one of the sixteen international agricultural Research Centers within the Consultative Group for International Agricultural Research (CGIAR). Its research mandate focuses on potato, sweet potato and Andean root and tuber crops systems, and natural resources management in high mountain areas, a reason why the CIP-FFS projects in Uganda were based on potato and sweet potato. The mandate of CGIAR is to develop technologies that contribute to poverty alleviation and food security by enhancing agricultural productivity. The choice by CIP to use participatory methodologies was based on the need to build platforms that enable closer interaction of the researchers with farmers, hence revival of a participatory research approach that was seemingly getting lost in the 1990s (Thiele *et al.*, 2001). With the major objectives of disseminating improved varieties, conducting participatory varietal trials and evaluation, and generating technologies that are responsive to farmers' needs, CIP used FFS in several of its research and development activities relating to potato and sweet potato integrated pest, disease and crop management as a platform for participatory research and farmers' learning (see van de Fliert *et al.*, 2002). Under the proposal entitled 'Integrated management of potato late blight disease: refining and implementing local strategies through Farmer Field Schools' CIP secured funding from IFAD in 1999 (Kai-yun and Yi, 2001). Through country offices in the different regions and countries CIP linked up with National Agricultural Research Systems for implementation. Activities of late blight management were then expanded to Uganda, in addition to Bangladesh, China and Ethiopia. In Uganda, the CIP project on potato was implemented in Western Uganda (Kabale district) between 1999 and 2001.

In mid 2001, CPR-DFID called for the development of small promotional proposals, through which CIP with its partners secured another project on *'Promotion of sustainable sweet potato production and post-harvest management through farmer field schools in East Africa'*. This collaborative project between CIP, the Natural Resources Institute (NRI) UK,

---

[5] Others refer to i@mak.com as "innovations at Makerere committee". Irrespective of the derivation of its name the work of this committee was expected to lift the status of Makerere in areas of relevance to the realities of the community.

and the Global IPM Facility in Uganda and Kenya, was implemented for three years - from April 2002 - and wound up in March 2005 (Anonymous, 2003: 4). The project's purpose was specifically to increase the returns from sweet potato enterprise through enhancing East African smallholders' capacity in sustainable production and post-harvest management. The project aimed to expand sweet potato Integrated Production and Post Harvest Management (IPPHM) throughout East Africa, and build on and share the lessons of the FFS process. This fed a more general purpose, espoused by CPR-DFID, of promoting strategies to reduce the impact of pests in herbaceous crops in forest agriculture systems in order to improve the livelihoods of poor people (Stathers et al., 2006: 4). At the initial stages the project was to focus on production, pest management and marketing but later broadened to cater for post harvest handling, storage and processing for value addition to sweet potato, hence IPPHM. Inclusion of post harvest handling and value addition packages was to encourage commercialisation of the crop. The need to commercialise the sweet potato enterprise, however, was neither brought up by farmers themselves, nor was it as a result of community analysis on the part of CIP. It was based on information from extension workers who worked with the FAO-FFS project in the area before. One questions how realistic and relevant marketing of sweet potato was to the local context, given the source of the information. The farmers in the project operation area cultivate sweet potatoes mainly as a food crop and not for commercial purposes.

2.2.3 The Food and Agriculture Organisation (FAO)

The role played by FAO in FFS is reflected in two main activities. First, FAO has played an important role among international agencies that set technical standards and influence the global policy framework on IPM. FAO conducted field projects mainly through collaboration with Global IPM Facility, a separate entity funded by the World Bank and three UN organisations (FAO, United Nations Environment Program (UNEP), and the United Nations Development Program (UNDP)). The Global IPM facility's mission is to promote the adoption of IPM, specifically based on the Asian experience, to farmers, governments and NGOs throughout the developing world (Sorby et al., 2003). Since its Asian Inter-country IPM programs from the late 1980s on rice, cotton, and vegetables, FAO has promoted and continue to consider Farmer Field School (FFS) as one of the most promising and successful approaches to promote participatory IPM. The Global IPM facility, too, emphasizes the FFS model/approach for farmer training in IPM. The FFS methodology now covers more than 76 countries across the globe with topics ranging from crop, livestock, soil and land, and health (Braun et al., 2006). This shows how FAO is determined and committed to spread use of FFS as much as possible. The different technologies or topics were used as means through which the model/approach was used.

Second, FAO is the agency responsible for international conventions, especially the International Code of Conduct on Distribution and Use of Pesticide, and other technical guidance on pesticide use (see Eddleston et al., 2002). IPM, broadly endorsed by crop protection disciplines as a combination of techniques to constrain pest development with minimal or judicious use of pesticides (Kogan 1998; Matteson, 2000; Gurr et al., 2003), has a central position in the Code of Conduct. Focus on reduced use of pesticides was mainly due to adverse health, environmental and economic effects, as clearly stipulated by Jeyaratnam, (1990), Micklitz, (2000) and Ecobichon (2001) among others. Pesticide in the broad sense refers to any substance or mixture of substances intended to prevent, destroy, or control any pest in crop and animal production. Pests broadly include vectors of human and animal disease, unwanted species of plants and animals causing harm or otherwise interfering with the production, processing, storage and marketing of agricultural commodities (FAO, 1986). In the IPM context, however, pesticides mainly refer to synthetic chemicals used, especially in the field, to control (insect) pests for increased agricultural yields.

With the overall goal of assessing effectiveness of FFS in addressing poverty issues FAO implemented a project, the East Africa Sub-Regional Pilot Project for Farmer Field Schools, operational from September 1999 to mid-2002 under IFAD and Global IPM facility support. FFS, as an extension model, was used in Uganda with the objective of building the competence in the (public) extension system, in responding more effectively to farmers' knowledge, information and technological needs, hence reflecting local conditions (Okoth *et al.*, 2002). FAO collaborated with other institutions for the implementation of the project: the National Agriculture Research Organisation (NARO) to identify the most suitable crop and site, and the district agricultural extension system for fieldworkers for day-to-day operation of FFS. Sourcing of IPM-FFS experts (from Zimbabwe) to train extension workers in East Africa as core/national facilitators, technical backstopping, and setting the training agenda were done by FAO. Use of experts was necessary in training the extension workers about the operation of FFS, since it was a new model. The training, however, did not give much if any chance to evolve an FFS suited to Ugandan (or East African) contexts. Telling facilitators what to do, based on theory generated in Asia, without encouraging and nurturing its evolution through practice reflecting local realities was not far from technology transfer.

At the initial stages of implementation, IPM did not fit the need for an entry point, because farmers did not use a lot of pesticides as had been anticipated. With a perceived background of frequent pesticide use in vegetable (tomato) and cotton production, based on experiences elsewhere, FAO-IPPM came in with a focus on IPM to emphasize the importance of reduced pesticide use. Although pests and diseases remain one of the major constraints in agricultural production in Uganda, agro-chemicals were not as heavily used by farmers in the area as in Asia; where (as mentioned) IPM-FFS began as a necessity. The proposed IPM project was then transformed and broadened into an Integrated Production and Pest Management (IPPM) scheme to cater for the whole crop production cycle from field through storage to marketing. Crop husbandry practices, post harvest handling and marketing were among the most important general problems that affect farmers besides pests. The operationalization of IPPM, however, was limited to crop production and protection. Marketing, a priority problem in the cotton production system, was not addressed. Changing the labelling of the technology package (from IPM to IPPM) did not cause any major changes in the operation of the FAO programme. Practically, IPM remained the technology package promoted, with a message of reduced pesticide use.

Understanding and caring about the real situation of farmers at the lower level remains a prerequisite for choice of appropriate technologies suited to targeted local sites or contexts. Transforming IPM into IPPM allowed FAO to position an already designed project to fit the realities of minimal pesticide use. Change at the implementation stage suggests that FAO made little effort to explore the applicability of the basic (IPM oriented) technology approach in the Ugandan context and therefore did not understand the practices of the 'community of learners' in the target areas. The thought of training government extension staff in IPM-FFS led to the view that FFS was the only or best way to teach farmers or disseminate technologies. It was actually one of several ways of working with farmers and technology. The training, however, neither equipped extension workers (facilitators) with adequate community analytical skills, nor new functional organisational skills needed to promote different and effective interactions with farmers around new and suitable technologies. Evidence will be discussed in later chapters.

2.3 Actors at the national level and their influence on FFS operation

At national level actors involved in project implementation included staff of the Ministry of Agriculture, Animal Industries and Fisheries (MAAIF); the National Agricultural Research Organisation (NARO) and Makerere University. The role of the ministry was mainly to give approval for the various projects in the country. NARO and Makerere researchers identified and supplied "on the shelf" technologies from their organisations. The priority attached to a technology component was based on its appropriateness in promoting the objective of the researchers at international level. For example promoting sweet potatoes remains very appropriate in meeting the mandate of CIP and promoting reduced pesticide use remains very appropriate for FAO (as earlier discussed in 2.2.2 and 2.2.3), which in turn are suitable in addressing the objectives of the funding agencies (see earlier section 2.2.1). This suggests that even the home research organisations were little prepared to find out the suitability of the selected technologies for the targeted farmers. Actors at the national level, in effect, used FFS to reinforce the mandates of their organisations but were less concerned to take care of farmer's problems in the project areas. Part of the explanation is the mindset that technology answers farmers' problems. Further evidence will be discussed in chapter three and four.

2.3.1 Ministry of Agriculture, Animal Industries and Fisheries

Through the Plan for Modernisation of Agriculture (PMA) the Ministry of Agriculture Animal Industries and Fisheries (MAAIF) played a vital role in making decisions about which agricultural projects were to be approved for implementation in the country. Compliance with PMA and PEAP was the guide used for project approval. NARO, and the agricultural extension service were (and remain) under MAAIF. There was/is no Research-Extension Liaison Committee (RELC) at the national level. In the process of approving projects for PMA/PEAP compliance, PMA secretariat (informally) acted as a 'liaison desk' specifically for project authors/writers and their funding agencies. There used to be Research-Extension Liaison units in all research institutes (in NARO) under outreach programme. With time, however, especially with inadequate funding and establishment of Agricultural Research Development Centres (ARDCs), the Research-Extension units became redundant and phased out. To maintain connection with the communities and respond to community needs/priorities, outreach partnership initiatives were established at ARDC or zone level (NARO 2001). Under this new development, there is a zonal steering committee with (four) farmer representatives as members (out of 13 members on the committee) - as to whether the farmers on the zonal steering committee actively played their part in influencing projects in the ARDC is another story that this thesis does not venture into. At the national level (PMA steering committee) farmers were neither represented nor engaged. Absence of farmer representation in the decision making process concerning agricultural projects at the (Uganda) national level is evidence of a big disconnect within a system of reform emphasizing decentralisation, democratisation, stakeholder participation and empowerment of people at the lower level.

In Uganda, like many other countries in Africa and the rest of the developing world, development initiatives, and therefore projects, were/are funded by external agencies, which provide over 50% of government expenditure. Much of the expenditure is directed towards non-tradable goods and services like poverty eradication (see Atingi-Ego, 2005). Poverty eradication is multi faceted, though in most cases applicable to agriculture-related fields, given that most of the poor (96%) live in the rural areas, with a majority (84%) depending on agriculture for their livelihood (PRSP, 2001). But in the national annual budget allocations to date, however, the government of Uganda spends least of its own funds on agriculture, reasoning the sector receives significant donor project funding. All projects contribute to PEAP/PMA objectives, via increase in income and improvement of quality of life of the poor, and improved household food and nutritional security, directly and/or through the market, provision of gainful employment and sustainable use and management of natural resources. MAAIF assumes the role of streamlining non-government projects to comply with PEAP. These projects (as just noted) greatly complement the government's expenditure in the agriculture sector and add considerably to improvement of rural livelihoods.

Decisions about what problem to address in a community through development projects are made by the individuals, organizations or institutions who articulate the problem in a proposal, subject to government approval. Irrespective of the sector, decisions about which projects to accept or reject at national level are passed by highly-placed civil servants through a series of committees, following specific guidelines (MAAIF and MFPED, 2003). Project/program proposals developed by any agency are presented to the director of PMA who forwards them to the PMA projects/programmes sub-committee, who in turn, after scrutiny, forward them to the PMA steering committee for action on recommendations made by sub-committee. The PMA steering and sub-committees assess projects for compliance and conformity with PMA objectives. Marks (in percentage form) are awarded to proposals following a scheme designed by the PMA steering committee based on the objectives of PMA (see Box 1).

The formula used in assigning marks is subjective in regards to the way the committee understood the proposal, but not according to any objective, measured criteria (based on what

---

Box 1: Criteria for assessing a project proposal for PMA compliance.

1. Contribution to PMA objectives (70%)
- Increase in incomes and improvement of quality of life of the poor (30%)
- Household food and nutrition security directly or through market (10%)
- Provision of gainful employment (15%)
- Sustainable use and management of natural resources (15%)

2. Implementation within the PMA policy framework and principles (30%)
- Provision for multi-stakeholder participation in initiation, design, implementation, monitoring and evaluation (15%)
- Feasibility of project/program to be funded under PMA arrangements (15%)

---

goes on in the field). Based on this criteria the PMA project/programmes sub-committee reviewed 44 projects in 2003/04 (PMA 2004) and about 45 in the financial year 2004/05 (PMA 2005). Because many proposals did not adhere to the PMA compliance guidelines, rejection was found less helpful. Instead, the alternative of re-submitting after making required modifications was taken up. It then remained the choice of the author to pursue the modifications and resubmit (process of reviewing proposals is shown in Box 2).

The criteria, devised by PMA steering committee, seem vague and inherently unmeasurable - perhaps just based on gut feelings. There is need for a more objective scoring system that looks into the realities in the field while approving project proposals. True democratic farmer-

---

Box 2: Process of reviewing programmes or projects for PMA compliance.

In the absence of a director of economic planning, either the director of budget or any other commissioner on the committee chaired the meeting. The chair had the last word. NAADS and PMA were represented with specific interest given that they were key programs in the country under agriculture. Members on the committee held high positions (commissioners, principal economists, deans, principal community officers, principal policy analysts, rural development advisers, first counsellor and program manager/officer). Following a criteria developed by the same committee, members assessed compliance with overall national development objectives through awarding marks. Each member was given a copy of a 'marking guide' (see Annex 2) that stipulated the maximum marks each item (objectives and principles) scored. A decision to accept or reject a project/program was based on the average total mark/scored awarded, the pass mark being 60% and above, with action depending on degree of realignment. Absence of issues that required re-alignment led to approval while presence or need for realignment, focus or redesigning led to deferred approval of a project/program until it was realigned to the satisfaction of the committee. A score of less than 60% implied the proposal was not PMA compliant/compatible and was rejected.

---

devised proposal as well as farmers' active participation in developing criteria for approval of a community development project would not only make farmers to actively identify with or own the project but would also strengthen farmers in holding project implementers and themselves accountable. From the steering committee, recommended proposals are forwarded to the development committee for assessment in terms of value for money and conformity with Medium Term Expenditure Framework (MTEF[6]) for macro-economic stability. Some proposals have special treatment compared to others[7]. These include those developed to manage emergencies, those with ready funding, and those where donors were present in the committee, those based on cooperative working relationships with line ministries[8] and those especially well written/presented. For officially accepted project/program proposals, the ministry of finance, through its permanent secretary, approves and writes to proposed funding agencies giving them a go-ahead. The process seems to be more academic than practical. Most significantly, it does not include any representation by or consultation with targeted local communities at any level as indicated by the process and PMA steering committee membership.

PMA steering committee members (who are mainly urbanists in form of high level civil servants, NGO staff and scientists), are selected by senior civil servants or administrators of their institutions as requested by the PMA secretariat. They were mainly 'office people', in fact, with no call to go to the field to assess realities on the ground. Members included commissioners of planning in ministries of agriculture (Ministry of Agriculture, Animal Industry and Fisheries - MAAIF) and finance (Ministry of Finance, Economic Development and Planning - MFEDP), assistant commissioner (Ministry of Water, Land and Environment), principal economists (MFEDP), a university dean (Makerere University), a director of NARO who often co-opted one projects officer from NARO, program or project officers (Danish Development Agency and PMA secretariat), principal policy analyst (Ministry of Local Government), principal community officer (Ministry of Gender, Labour and Social Development), coordinator agriculture education (Ministry of Education and Sports), first counsellor (European Union secretariat, not Ugandan national), rural development advisor (World Bank) and assistant resident representative (FAO). Occupied mainly with administrative work, given their positions, they had the freedom to co-opt other persons as appropriate. Although it saved time and probably resources that might otherwise have been involved in selecting mandated representative, a clear problem with the present system of nominees from nearby is that it is unclear what constituencies and interests the persons on the committee are supposed to represent. On enquiry, it turned out colleagues in the same institutions did not know about the committee, the involvement of their institution in the PMA committee, or indeed what the committee was about!

The marking and assessment process was bureaucratic and mechanical, with no clear means to assess, in practical terms, what might work and how, let alone to assess whether it met local

---

[6] MTEF is an annual rolling three-year expenditure planning. It sets out medium-term expenditure priorities and hard budget constraints against which sector plans can be developed and refined.

[7] This was the assessment of one of the members of the committee.

[8] Agencies that developed projects in the field of agriculture were expected to work with the ministry of agriculture because it was responsible for agriculture in the entire country.

needs. The criteria of how well a proposal was written or presented apparently counting for more than whether it could be beneficially implemented seems inappropriate. It seems only likely to select for skill in writing good quality proposals and to encourage reliance upon this skill as an income generating activity in the name of community development. Decisions about what was good for the communities were made by bosses at high/national level without consultation with or involving the affected and targeted people at the lower level. Absence of a verification procedure to ensure whether elements or parameters in the marking guide actually correlated with anything meaningful at the implementation level ensured the process remained a purely ivory tower exercise. The fact that the committee had no real, developing view of what worked and what was needed, through on-the-ground assessment, only left researchers free to build pictures of farmers' alleged needs reflecting little more than the thinking and needs of their institutions.

A mechanism to include farmer representation[9] on the committee and to revisit all projects on ground for conformity with PMA would make the committee process more relevant and responsive to prevailing problems at community level. That members of the committee might have stakes in some projects cannot be ruled out. It was observed by one committee member that some proposals were more easily approved. This included those with ready funding (so as not to lose the opportunity, since most offers of funds are time bound); those from the director-general of NARO, and those where some "well-positioned" persons in the ministry of finance/government had vested local interests. Approvals of such proposals suggested operation around 'technical know-who' (personal relationship) instead of 'technical know-how' (competence required for performance). Approval based on personal relationship breeds inefficiency in delivery of effective community development and is a waste of resources.

This process of assessment for approval, however, did not apply to already registered NGOs in the country and projects implemented before 2001. FFS projects therefore did not go through this process because they were mainly implemented by existing and registered NGOs. Besides, they were smaller projects. This explained why the committees did not know much about FFS projects, save for the second phase FAO FFS project that was supposed to work with and fit in the NAADS framework. There was some conflict in the way the project implementers desired the implementation process to be, as opposed to what is desired under NAADS, especially in terms of farmer training approaches as well as monitoring mechanisms, among others. In NAADS, monitoring was carried out by farmer structures called farmer *fora* at parish and sub-county level, yet in FFS there is a desire to develop a participatory M&E framework with farmers in FFS groups. The negotiations around adjustments were on-going, however, since the implementation of FAO-project phase 2 began effectively early in 2006.

---

[9] In Uganda, there is a farmers' federation at the national level called Uganda National Farmers' Federation (UNFFE), with branches at district and structures at sub-county and parish levels. UNFFE's overriding objective is to mobilize farming community into one independent umbrella organization with a strong voice to lobby and advocate for farmers' interests.

2.3.2 Programme assistants: coordination and direction of projects

Interactions between facilitators (extension workers) and scientists went through programme assistants. Programme assistants, accountable to FFS project leaders, were stationed one per district, and served as representatives of the project team at local level. For programmes covering the East African region, like FAO-IPPM and CIP-IPPHM, programme assistants were recruited specifically to ensure smooth implementation of all day-to-day FFS activities and linkage with key stakeholders across the country. These appointments were made at national level and appointees coordinated FFS activities across the various districts within the participating countries. In this context, FAO-IPPM had three programme assistants - one in each participating country (Uganda, Kenya and Tanzania) - answerable to the programme officer at FAO head office. The CIP-IPPHM project, by contrast, had one programme assistant who coordinated FFS activities in all districts and countries (mainly Uganda and Kenya), and was answerable to the project leader, from NRI (a British organization). Programme assistants ensured that money and learning materials were despatched to the FFS groups, prepared field tours/visits, and ensured that facilitators met farmers as planned.

Because of prior involvement in FFS, FAO programme assistants and some facilitators provided technical backstopping to other organisations using FFS after 1999, especially during preparatory periods. This implies that FFS cadres (program assistants, district coordinators and facilitators especially extension staff) were competent in understanding technology practices and requirements in relation to local farming community contexts. Programme assistants were often called upon to provide information needed to develop FFS projects (even if for different organisations) more relevant to targeted communities. Although horizontal interactions and information sharing between FAO-FFS field staff and other FFS projects at national level encouraged better positioning of subsequent FFS projects in their local contexts, a 'teacher-student' relationship emerged nevertheless. FAO-IPPM-FFS field staff tended to assume the role of teacher (i.e. a person possessing the "right" FFS knowledge) and the other organisations acted as students. The teacher is assumed to know more and therefore exercises a power relationship over learners! The teachers, who were mainly extension workers in project coordination positions, rarely stayed with farming communities but resided in town[10]. The time they spent with farmers was rarely long enough to enable them get a clear understanding of the local agrarian situation. Staying with farmers is one thing and having the knowledge and skill to analyse what is going on is another. The extension workers (or facilitators) generally lacked the community analytical skills to enable them get a better understanding of local communities. Most of the information provided about local communities was more a matter of what extension workers perceived or assumed to be happening, rather than the result of sociological or ethnographic analysis. This account of how projects operated seems to indicate that a teaching mode was assumed (perhaps mainly for administrative reasons) at a point where there was a clear need to shift to a learning mode through which actors might interact

---

[10] Most extension workers preferred residing in town and just move to the rural when they were to work with farmers. In town they access social services like electricity, better housing, better education for their children and meet many friends as compared to rural whose social services are inadequate or even absent in some cases.

with the targeted communities directly, and thus gain a careful listener's awareness of what went on concerning real and pressing local issues.

2.3.3 Research institutions at national level

So far, a technographic account of programme set-up seems to suggest that farmers were not (so far) involved in problem articulation. This leads to a paradoxical possibility that participatory projects follow the logic of donor-led project design, thus excluding participation! Involvement of national institutions brought in another layer of "business as usual". If FFS could be seen to improve chances of farmers taking up new or improved technology it would be more likely to be judged a success. When national researchers joined the process they used available technologies on the shelves to fit packages required in IPM, IPPM and IPPHM. These technologies (as discussed in chapter three and four) basically included improved/new varieties and recommended agronomic practices. Technology components under cowpea and groundnut IPM for example involved improved varieties (MU-93 cowpea and Serenut groundnuts), recommended spacing and spray regimes, all of which had been developed earlier. Those under IPPHM included orange fleshed sweet potato varieties and recommended planting methods. Choice of specific technologies brought each FFS project closer to the usual ways in which agricultural research organizations worked with farmers. Provision of these technology components was mainly the job of national agricultural institutes.

National Agricultural Research Organization (NARO)
National Agricultural Research Institutes (NARIs) have to assume a role in any international effort targeting small-farmer agro-technology solutions, because they are the mandated organizations for in-country agriculture research. In Ugandan FFS the main player of this kind was NARO, the largest and most established national agricultural research institution. Related to NARO are the agriculture research development centres (ARDCs) where adaptive research for zonal agro-ecological needs is conducted. ARDCs support the outreach programs of NARO. Universities, too, operate as NARI. Makerere University - in particular the Faculty of Agriculture - became involved in FFS activities.

NARO was established by an act of Parliament, on 1st November 1992, with a mandate to undertake, promote and coordinate research in all aspects of crops, fisheries, forestry and livestock, and ensuring dissemination and application of research results. Note that dissemination of technologies is not within the mandate of NARO. Performance of research organisations (at this level) was often measured in number of technologies produced in a given period of time, but not necessarily in terms of their uptake or relevance to community problems. An inventory of technologies (NARO, 2002) was made after ten years of operation. The assumption was that the greater the number of technologies the more NARO had contributed to rural development, through making improved technologies available. For a technology to contribute to rural development, it should be useful to the intended user in solving farming related problems. In developing improved or new technologies (varieties), scientists consider yield and resistance to pests as the principal characteristics yet farmers look at taste (edibility) and cost as the main factors (detailed evidence is shown in chapter four). That means that what is supplied and what is demanded take different directions. Failure to identify what farmers need leads to a tendency of developing technologies that only end up on the shelves. To develop technologies from any research institute (NARO) that are useful to the beneficiaries, a tag of demand from the users is necessary. The demand tag motivates uptake and usefulness of the technology. Whether such endorsement has been forthcoming is a matter that can be debated.

Development and dissemination of technologies is done at research institute level through different programmes. NARO is an umbrella organization resulting from an amalgamation of nine research institutions that existed within several Government Ministries. The nine research institutes have varied but complementary research mandates. Linkage of research institutions at the international level with NARO was mainly through research institutes that hosted programmes on specific technologies or crops that matched the interests of the scientists at international level. These programmes are not very directly linked to the needs of Ugandan farmers, nor have they (with rare exceptions) been shaped by direct consultation with local farmers. This is the general trend for agricultural projects originating from outside the country. The role of national research staff was mainly restricted to testing and adaptation of techniques or improved planting materials from elsewhere, and making these improved technologies available locally. In effect, this role was continued into the Ugandan variant of FFS. Scientists provided technical backstopping to field workers on recommended technologies, took a lead in training prospective facilitators, and in curriculum development, and prepared reports for the project. This not only served to boost the status of the national research organisations, but also helped impose a certain harmonisation of innovation styles on projects at the national level. This harmonisation tended to keep researchers at international and national level looking in the same way at technology as a solution to problems in society thereby strengthening the view that farmers are receivers and users of technology, not initiators. FFS had no specific strategy - when introduced to Uganda - for breaking through what the anthropologist Mary Douglas terms an institutional "style of thinking" (Douglas, 1996).

In partnership with NARO Uganda, CIP re-established potato research and seed production in Kalengyere from 1989 to 1994 (Low, 1996). Kachwekano-Kalengyere ARDC was the station responsible for all research related to potato in the western agro-ecological zone/region of the country. For the case of FAO-IPPM, the first link was with the national cotton programme hosted at NARO's Serere research institute. CIP-IPPHM linked up with the national sweet potato program hosted at NARO's Namulonge research institute. International improved soil fertility projects, including the ISPI-A2N linked up with the NARO Kawanda research institute, where soil fertility related programmes are normally hosted. These programs had generated a variety of technologies from which selections were made that fitted with IPPM and IPPHM. When adapting "on-the-shelf" technology to local conditions researchers typically are used to on-farm research as a way of disseminating and evaluating technologies with farmer collaboration. This well-established approach tended to creep through the Ugandan FFS at the expense of farmer-based discovery learning. Linking up with FFS became a means to further disseminate already generated technologies to the farming community detailed evidence come in chapters three and four.

Under NARO, research is organized around themes from which specific programmess and projects targeting specific commodities or technologies are developed (Bashasha *et al.*, 2004). For example, under the root and tuber crop theme cassava and sweet potatoes belonged to separate programs. Under the sweet potato program, activities to enrich nutritional value through development of orange fleshed (OFSP) varieties, and diversify and fortify sweet potato based products commenced around 1998. To generate and disseminate improved varieties, the program collaborated with various organisations coordinating research activities at regional and international level. For example, the Regional Potato and Sweet Potato Improvement Program in Eastern and Central Africa (PRAPACE) coordinated and facilitated research on sweet potato and *Solanum* potato within the eastern African region, and the International Potato Center (CIP) coordinated and provided technical backstopping regionally, and facilitated germ plasm exchange (including supplying materials cleansed from viruses), in cooperation (locally) with the Ministry of Health, MAAIF, and various NGOs and community-building organizations. NARO, like any other research institute, had no mandate specifically to disseminate technologies; technology dissemination was the role of the extension service. But involvement in FFS was attractive in that it provided an opportunity for on-farm evaluation of technology through the integrated packages promoted in FFS projects. NARO did not run FFS on its own but was involved as an implementing partner. However, as explained, NARO was driven by a larger logic of international cooperation, within which it faithfully transferred its agreed role (adaptation of international materials and techniques) into the FFS arena, perhaps (as we shall see) at the expense of the "bottom-up" approach to innovation FFS was originally focused upon in the IPM field.

Makerere University (Faculty of Agriculture)

Makerere University as an academic institution is built on four pillars: training, research, outreach and partnerships with other actors in research and development. Graduates from the university (faculty of agriculture) are employed mainly in agricultural institutions, both private and public, where they fulfil a variety of different positions and roles. These institutions include extension, research, academia, MAAIF, industry and commercial farming, among others. The university participated in training FFS project staff both directly and indirectly. The indirect path was though its graduates employed in the various institutions involved in FFS implementation. The role played by actors (professionals) in FFS implementation reflected how they had been trained while at university, therefore. Again the issue of "institutional styles of thought" comes into view.

Research at the university has mainly been on-station experimentation, carried out either on the university farm or at one of the institutes of NARO. The trend among researchers is currently shifting towards on-farm experimental work, though with very few farmers, often in fact only one host farmer. Choice of area of research varies according to individual interests, project objectives, adequacy of facilities, and convenience. Most research activity carried out in the University aims at production of publications - either conference papers or journal articles, which count for promotion. Since researchers generally have an adaptive or applied orientation the approach does results in technologies, but not always suited to farmers' realities in the field. The problem is not unique to Uganda - the research station environment (e.g. with highly fertilised soils) and management practices (e.g. frequent watering or weeding by workers) is very different from that associated with farmers' fields, and it is common to find that research station results are hard or impossible to obtain in farmers' conditions. This is one reason for the current shift towards more on-farm research, but few on-farm experimenters yet study the farmer-technology interactions as closely as they study soil-climate-crop-pest-disease interactions. This (it has been argued) is one reason for the rather low productivity of agricultural research in Africa, in terms of uptake of technologies relevant and applicable to farmer conditions (Richards, 1985).

In recent years sponsors of research have advocated and supported adaptive or applied research in the hope of more directly contributing to improvement of typical farms and farmers in Africa, and research at Makerere (Faculty of Agriculture) has not been exempted from this shift towards funding of research directly applicable to farming livelihoods. As a result, research has paid greater attention to the context in which African peasant farmers operate. But in spite of efforts actively to involve farmers, researchers are still trapped by a system where even on-farm research output is measured more by publication criteria than increased productivity and success among farmers (Kibwika, 2006). The actual problems to be addressed are still rooted in disciplinary discourses, whether the Green Revolution search for "super" crops, or analytic methodologies such as production systems ecology. In either case, the agenda tends still to be set by the researcher, with the assumption that once results are "tested" in terms of acceptability of findings to the editors of scientific journals they amount to a technology ready to be disseminated to the farmers as recommended practices. A certain rigidity of thinking reproduces itself across generations of researchers, according to the orientation of training received and the institutional mindset of researchers. What might suit the local is still handled at the level of basic assumptions guiding the research, and more rarely as a researchable objective in itself. This is sometimes referred to as the unresolved problem of "pre-analytic choices". It seems likely (as will be demonstrated in later chapters) that these choices are more firmly grounded in what suits the research organization than in a serious concern to solve farmers' prevailing problems.

An example illustrates the point about the institution (in this case normal science) doing the thinking (Douglas, 1986). In response to proposal calls for research relevant and practically useful to the farming community, two FFS projects, among others were developed. Under a Rockefeller Foundation Grant, a cowpea improvement project commencing in the early 1990s developed IPM components to minimise costs of production incurred by cowpea farmers as a result of frequently using chemical pesticides as the sole remedy for devastating pest damage (Karungi et al., 2000; Adipala et al., 2001). But use of pesticide was a practice mainly associated with commercial cowpea farmers (Isubikalu et al., 1999). Inadequate funds did not permit dissemination of the developed IPM package to the wider cowpea growing community for whom it was designed. Following airing of the charge that the universities were not responsive and active in research relevant to needs of the society (Patel and Woomer,

2000), an initiative known as "innovations at Makerere for the community", abbreviated as i@mak.com, was funded by a sub-Saharan African consortium comprising the Rockefeller Foundation, Carnegie Corporation, and World Bank, calling for proposals from university staff to develop projects to reach out to rural communities. This was an opportunity to develop a project to disseminate the formerly developed cowpea IPM package to the intended communities.

Sharing and feeding-back research results to the farming community necessitated a participatory approach that involved more farmers. But in the event, on-farm research was largely researcher-driven and managed. Only one farmer hosted the experiments. To effectively reach out with developed IPM package to more farmers, FFS was then chosen as the most appropriate model (Karungi and Adipala, 2004), building on experiences from Asia in training farmers about IPM in rice and other crops. The aim of using FFS in cowpea production included: (1) introducing improved cowpea/groundnut varieties as better alternatives in pest management (2) having farmers involved in making appropriate choices of practices better fitted to their individual local contexts, and (3) demonstrating the economics of more judicious pesticide use. The project was implemented in the main cowpea and groundnut producing districts in Uganda (Kumi and Iganga). Groundnut FFSs were established in Iganga because (a) it was the leading producer of groundnuts, and cultivated the crop twice a year, and (b) it was a hot spot for rosette, the most important disease of groundnuts. Kumi-Bukedea hosted a cowpea FFS because (a) it had the highest number of cowpea commercial farmers using large amounts of chemical pesticides, and (b) it was one of districts where IPM technologies had already been developed through previous on-farm research.

Through i@mak.com support, a project for safe pesticide use and handling (SPUH) was funded. Although the levels of pesticide use in Uganda on the whole were low, there were cases of farmers abusing the use of pesticides through ignorance. There was concern that continued and cumulative pesticide abuse would lead to environmental and human health hazards in the near future. The project was in effect a preventive measure in safe pesticide use. It was in this context that the project was initiated with specific objectives. One was to teach farmers the safest way of handling and using agro-chemicals, while minding about their lives. Lessons included the ideas that chemicals are poison to people, and have an effect on male fertility (perhaps leading to impotence). A second aim was to encourage conservation of the environment and maintenance of balance in the ecosystem. Lessons included safe disposal of unused pesticides to avoid contaminating water sources and destruction of beneficial insects like bees. A third aim concerned the need to reduce amounts of chemical residues on crops required on the international market. Countries like Uganda, low in dependence on pesticides, have a possible advantage in seeking to supply the increasingly demanded organically produced crops on the global market. The idea was to make farmers aware of this opportunity and get them to take the issue of chemical pesticide use and abuse seriously.

Proper use and handling of pesticides had links with IPM strategy to limit exposure of humans and environment to pesticides. It was against this background that the project was seen to have practicability and relevance for major vegetable crops, among potential study crops where frequency of pesticide use was perceived to be high. Tomatoes and cabbage were the major vegetables grown in Mukono and Kiboga districts where the project operated. The choice of whether to take on tomatoes or cabbage devolved upon the FFS group, although even here the actual choice was strongly influenced by the researchers and facilitators. For example, in one FFS, the facilitator was more informed about tomatoes given that this crop was the subject of his M.Sc. degree, and unsurprisingly it was chosen. Members may have reasoned quite rationally that if they were to give up time to FFS events they might as well pick the theme that the facilitator knew most about. But it may not have been (perhaps almost certainly was not) the theme farmers would have picked if it had really come down to their choice alone. In another case, cattle was chosen as the study animal, not because a majority of farmers had cattle, but because it is the highest value item among livestock, and therefore would be likely to attract many farmers who dreamed of owning cattle.

Scientists at national and international levels used FFS as a form of on-farm research to induce farmers to fit in with what they had on offer, in the hope of evaluating and disseminate their already generated research findings, and not (as it was intended) as a means to understand local contexts in order to develop more appropriate technology. The researchers' strategy of having as many farmers involved in FFS projects as possible was likely intended as a smokescreen to cover over lack of active engagement of farmers in identifying priority needs at the lowest level. But this was no bad thing in researchers' eyes, since they reflected a tradition of thinking in which poverty was seen as a condition defined by lack of technology. Inject technology and all should be well. Kibwika (2006) has traced its roots in Uganda back to the colonial period. The above short example of FFS failing to throw of the shackles of a transfer of technology approach, despite donor enthusiasm for "participation" seems to imply that dubious assumptions about technology "came with the terrain" (i.e. they were part of the culture of the researchers and institutions through which FFS was first applied).

2.4 Local actors and their influence on FFS operations

Local level actors are found at the district, sub-county, village and village level, mainly associated with community mobilization for the projects. Major actors included the agents of the agricultural extension system at district level and members of the farming community at village level.

2.4.1 Agricultural Extension service delivery and roles played

A multitude of institutions and organizations engage in information dissemination with different aims and emphasis in rural Uganda. Following both public and private extension service workers partly reveals the areas of emphasis of the different FFS projects. FAO worked exclusively with government extension workers, because it was feasible to introduce and therefore spread use of FFS as an alternative or improved extension service delivery mechanism. CIP on the other hand engaged NGOs, in addition to the government (or public) extension service, to spread the potato and sweet potato technologies to more farmers. Makerere projects worked closely with government extension workers, in part because they were a more appropriate link to engage the local government leaders in and legitimate project activities hence, and also because government extension workers are more readily available to and accommodating of the interests of workers from other public institutions. It is easier to establish, and maintain, linkages with public service systems than with private service systems, which are in most cases short-lived (it is all too easy to lose the contact person for a short-lived project).. Institutionalization of a new idea is easier when working with an enduring public institution/organization. In Turkey, Ozcatalbas *et al.* (2004) observed a situation comparable to Uganda in which international institutes mainly cooperated with public institutions (while also concentrating, as we found above, on disseminating existing information rather than producing new information).

Public extension service

Technology components of IPM, such as new/improved varieties, were developed under conditions that must be observed to determine whether they yield desired ultimate returns. Because of the way such technologies were developed, farmers required to be taught about the conditions to use these new technologies effectively. Public extension workers are the main actors in disseminating (new) technology (Turrall *et al.*, 2002; Tesfaye, 2005), mainly as recommended or better practices. Verbal messaging is the major form of information dissemination used in public extension. In Uganda, public agricultural extension services have been decentralized at the district level, with all agricultural extension field workers answerable to the District Extension Coordinator (DEC) who assigns any additional task to the extension workers. Under this decentralized arrangement, extension workers were placed in charge of all production related activities within the sub-county of their jurisdiction. Any projects that contained agriculture-related activities in a sub-county were supposed to link up with the sub-county agriculture extension worker. Projects operated in small areas (few farmers, in few villages, in few selected parishes). Government extension worker, by contrast, are responsible for advising and guiding all farmers, both individually and in groups, throughout an entire sub-county. This is a very challenging task given the very low ratio of extension workers to farm families (about 1:500). Often, extension workers lack transport to reach many of the farmers, not to mention kits for demonstrations or stationery to keep records and file reports. As a result, extension agents were not frequent visitors in many villages, and many farmers did not see them from one year to another. Given limited resources most agents preferred to concentrate on particular hardworking groups within ready access. Anderson and Feder (2004) mention the large scale of operation as among the major challenges facing public extension.

The extension service in Uganda is mandated to teach farmers about new/improved agricultural technologies with the aim of improving (or modernizing) farming. Extension workers, therefore, are supposed to link with research and development organizations to acquire new/improved technologies to pass on to farmers. Normal instruction is based on what extension workers perceive as the needs of farmers (i.e. it is supply driven) and only rarely on what farmers demanded to learn about (demand driven). Many times, extension agents would limit themselves to what they learnt about while in the university or college, to ease their work load. In what effect amounted to informal lectures, based on how extension agents had themselves been taught (Kibwika 2006) farmers were told the 'right' thing to do. This method assumed that either farmers did not know what they wanted (as extension workers and researchers often asserted) or that what they knew was inadequate. In effect, extension staff failed to analyse the feasibility of what they taught and how well it was received. Actually, farmers were generally not keen to attend meetings called by extension staff because they seemed to offer no 'news'. This prompted the preference to work with farmers in groups. Group dynamics kept farmers engaged to some extent. Troublesome or disenchanted farmers, who had already "voted with their feet" by not joining a group (i.e. the vast majority of ordinary farmers, (cf. Bukenya, 2007), could be safely ignored.

Demonstration sites were mainly used by NGOs. On a few occasions private organizations, like Sasakawa Global 2000 (SG2000), provided demonstration materials to public extension workers working within specific sub-counties. It was observed that in general it was difficult for NGOs to liaise with other players, such as the government extension service, since this would mean adopting the delivery models and technologies of other organizations. NGOs have their own mandates and funding mobilization strategies, and this generally involves having a distinctive purpose, and associated technology packages, delivery methods, time frames and areas of operation. By and large NGOs work with the chosen few, as opposed to public extension, which is open (in theory) to any one, even if in the worst case it works with no one due to demoralization and lack of funds. The rigidity of NGO time frames also encourages a focus on technology with minimal attention to farmers, because the presence of an "alien thing" can quickly be counted by assessors, and its visibility convinces donors that change is afoot. So often, the appearance of change is carefully massaged to satisfy evaluation requirements, with artefacts barely surviving in use beyond the end date of the project. But this emphasis on short-term visible effects, undesirable in itself, also tends to devalue or undermine less tangible aspects of extension (e.g. timely advice, training farmers in a skill, or fixing broken equipment).

More is said about private extension in its own right in a section below. Here, it is relevant mainly to note that public extension workers have a tough job, given resourcing constraints, made tougher by NGOs "show-boating" alongside. In Uganda, as in many parts of Africa, government extension services tend these days to be used as a pool of skilled labour to be assigned to work with any organization or institution that needs their service, whether on formal or informal terms of cooperation (including ad hoc activities at an individual level). Formally, organizations link up with the DEC, and the DEC assigns projects to extension workers, and therefore to villages/areas of operation. The problem with this is that it reproduces existing institutional habits.

Under PMA[11], extension workers were provided with motor-cycles. Maintenance and fuel costs were left to individual motor cycle owners, i.e. extension staff. Projects like FFS used existing government agricultural extension workers, as facilitators, to implement their activities. Such projects provided training materials and transport (fuel allowances) to enable

---

[11] It was around 1999-2000, through the PMA programme that university graduates from agricultural science faculties (Agriculture, Forestry and nature conservation, Botany, Zoology, Fisheries and Veterinary Medicine) were first employed as extension workers at sub-county level.

extension staff to work more efficiently. Transport incentives from projects motivated staff and ensured ease of reaching many farmers in a shorter time than before. There was then a tendency not to care about the majority of farmers who did not have access to FFS projects. An apparent danger is that FFS projects served as ventures creating islands of NGO-style commitment in a more general sea of neglect, encouraging the marginalization of the majority. Sometimes some extension workers played neither their role in central government nor in projects. They kept using one side as an excuse for not playing their role on the other side. While not at work, most extension workers preferred spending their time with friends and families in town - where they could access better facilities like medication, education, electricity, and communication, among others. This partly explains why most of them did not stay at the sub-county headquarters.

In Uganda, provision of government extension services has largely been a supply side venture. Very few farmers sought advice from agricultural officers. It was the extension staff who drew up a program and looked for farmers. The same trend continued into the FFS period, even though the new NAADS programme was meant to build a demand driven type of extension (Bukenya 2007). The NAADS program started in 2001 with a few pilot sub-counties in a few districts, but has continued to roll out to new districts and sub-counties since. The essence of NAADS was to create the beginnings of a market in extension by setting up a system whereby farmers drew on advice and inputs via contracted service providers (Bukenya 2007). But apparently, most service providers under NAADS were not technically competent and often hired extension staff to carry out necessary training. Subsistence allowances paid under NAADS were attractive, and kept most extension workers busy either in their sub-counties or in other sub-counties as requested by a service provider. Yet, rather paradoxically, extension workers were not expected to become service providers themselves under NAADS, so long as they remained government workers. It is yet rather unclear what will become of public extension under these new circumstances, except to say that projects such as NAADS and FFS seem to set up competing, and perhaps contradictory demands. The present picture is a patchwork quilt of nodes of activity in a wider sea of neglect. Whether the nodes take resources away from the majority (i.e. whether there is now active exclusion where before there was indifference) and whether this will feed rural discontent is as yet hard to decide.

Private (NGO) agricultural extension services

NGOs play a major role in extension and education services in communities at the grass root level, though covering (as noted) only favoured areas and not the entire client base. Nightingale and Pindus (1997) associate NGOs or private extension service delivery with increased flexibility, resulting from reduced bureaucratic procedures. In Uganda, NGOs promote specific technologies/services within specific geographical areas. For the purposes of this technographic sketch, we can gain an insight into the contribution of private NGOs by examining a few such NGOs before briefly also sketching the role of true private service providers (e.g. traders in agricultural inputs)

Africare is a non charitable international organization founded in 1970 as an agency focused on delivering health services. It first supported Uganda in 1979 -1981 with a mixed programme of medical and agricultural services. During years of chronic instability it withdrew, but returned to the country in 1996, with three major programs: resettlement and retraining of demobilized military personnel, food security, and health. In Uganda, the organization is currently operational in Western Uganda covering five districts (Kabarole, Rukungiri, Kanungu, Kisoro and Kabale) with a variety of projects, including farmer training. This last included a project specifically on farmer training in improved potato production, supported by IFAD (Africare, 2003:29). It was implemented in Kabale where potato is a major cash and food crop. Through this project, clean and improved potato varieties were secured from Kalengyere research institute and distributed to the farming community through groups formed to work with Africare. The link between Africare and CIP in this context was Kalengyere. Africare introduced the CIP-FFS project to its existing groups, and thus ensured that the two organizations (CIP and Africare) supplemented each other in achieving the objective of sustaining potato production. While Africare used demonstration plots to expose potato farmers to better/recommended agronomic practices for improved yields, CIP FFS used experiential learning with an emphasis on integrated management of potato bacterial blight and wilt diseases. Note that IFAD supported both projects. The link with Africare and the Soroti Catholic Diocese Integrated Development Organisation (SOCADIDO) perhaps reflects an emphasis in CIP-related projects to show results within the shortest time possible in spreading improved technologies to a wider community. Once again it is clear that Ugandan FFS fits within a technology transfer paradigm, with only lip service paid to discovery learning.

In Soroti district, a Catholic organization, *SOCADIDO*, was the first to work with farmers in groups in 1996. Its major clientele (and entry point) was Catholic women's groups in Soroti district, though the organization is accommodative of other faiths within the entire Teso region (mainly Kumi and Katakwi) with aim of supporting the farming community with services. Through its women's groups, and support for individual farmers, SOCADIDO implemented a range of activities with other development institutions/organizations, particularly those related to agriculture (CRF, 2001; SOCADIDO, 2001). This enriched its activities and contacts with the farming community. Activities in which SOCADIDO has been involved include agro-forestry, water and sanitation, HIV/AIDS prevention awareness and counselling, poverty alleviation awareness, group formation and development, literacy and adult basic education, sustainable agriculture, provision of farm inputs (pangas, sickles, wheel barrows, hoes, ox-ploughs and oxen), revolving funds in kind[12] (for heifers and goats), and monitoring of projects. Through the many groups established in Kumi and Soroti districts, SOCADIDO also carried out multiplication and dissemination of new and improved sweet potato varieties to a wider farming community.

Africa 2000 Network (A2N). *This* started in 1990 as a UNDP project but later (in January 2001) registered as an independent Ugandan non-governmental organization (NGO). A2N-Uganda has worked in Tororo since 1997 and trains community farming groups in development, gender sensitivity, organic agriculture, energy conservation, water harvesting,

---

[12] Every person given an animal (heifer or goat) paid back by giving an offspring of the animal taken to a neighbour. The cycle was repeated till every member was covered.

HIV/AIDS and health awareness and enterprise development. Tororo is one of the districts in eastern Uganda with very low soil fertility levels, recognized by the government of Uganda and other partners working in the area as an area of special need (Delve *et al.*, 2003). In an effort to promote appropriate soil management technologies for improved agricultural productivity, organizations working on soil related issues in Tororo, Mbale, Busia and Pallisa districts formed a consortium for collective effort. The consortium - Integrated Soil Productivity Improvement Initiative through Research and Education (INSPIRE) - was formed in late 1999, and included various NGOs (A2N, AT-Uganda, Kulika Charitable Trust, Plan-Uganda, CARITAS, SG2000, FOSEM/Cash-Farm), Makerere University, NARO, ICRAF & CIAT, as well as the Tororo local government. With support from the Rockefeller Foundation and FAO, INSPIRE used the FFS model as one of the methods for scaling up/out activities in which soil improvement and conservation technologies were tested and disseminated among farmers using experiments and demonstrations. Coordinating implementation of INSPIRE activities, A2N operated FFS in Tororo and Busia. The link between FAO and A2N reflected attempts by FAO to expand space to engage NGOs in the use of FFS model.

Agro-input dealers. This is an important if somewhat neglected group in the literature on African agricultural development, and it is one of the tasks of a technography to capture and integrate within a single descriptive frame all relevant aspects of an actual technological system. Private traders are more technologically active than often realised. They do not just supply agricultural inputs, but also at times disseminate important technical information, and sometimes teach farmers about available new technologies (crop varieties, fertilizers, equipment, pesticides). They are in business to provide technologies as demanded by farmers, and there is no reason why they should not be partner user groups in helping put into practice lessons from agriculture-related training where correct use of inputs is required (e.g. fertiliser use). Some agro-input dealers, however, abuse opportunities by supplying adulterated and fake technologies. In Uganda the majority are first and foremost businessmen and traders, not professionals. A recent study carried out by 3A Strategic Management Consultants (AT Uganda, 2004) to find out the distribution, composition, challenges and needs of agro-input dealers nation wide revealed a need for more knowledge and skills in agro-input handling. This was the reason why Makerere, under the SPUH project, invited agro-input dealers to a sensitization workshop and training on pesticide use. This enhanced their competence in extension services related to use of agro-inputs. Here we simply note that private input traders are currently quite important in Ugandan agriculture, though not yet significant for the poorest farmers, and that it is not impossible to envisage some interesting scope for public-private partnerships, perhaps based on applying FFS type discovery learning to the challenges of input dealing.

2.4.2 Local community leaders

The main elected community leaders are the local council chairpersons, mayors, councillors, and members of parliament all of whom are elected by adult suffrage[13] through voting at different constituency levels. The local council chairpersons (commonly referred to as LCs) are at village level (LC I), sub-county level and township (LC III), city and municipalities (mayors), and district level (LC V). These leaders are supported by councillors[14] who are also elected through voting by the same electoral colleges. In hierarchy, the LC I chairperson is answerable to the LC III chairperson who is also answerable to the LC V chairperson. The main duties of these leaders (commonly referred to as politicians) include mobilising the people for any development program, supervising or overseeing implementation of government programs or projects in their areas of jurisdiction and initiating community based development programs. Members of parliament are also community leaders (voted by constituencies that cover a maximum of 70,000 people) but are more of policy makers at the national level. Implementation is done by the local council leadership in collaboration with the technocrats.

Leaders at village level played a crucial role in mobilizing farmers, both in groups and as individuals, for FFS projects, as will be discussed in chapter three. They also officially opened and closed FFS ceremonies, such as inaugurations, field days, and graduation of FFS alumni, and in some rare cases monitoring some FFS activities when called upon by programme assistants or coordinators. Local leaders exercised some responsibility in ensuring coordinators played their community mobilization role satisfactorily. Assessment of performance by local leaders had nothing to do with the relevance or effectiveness of the introduced technology in the local context. Criteria focused more on whether project staff worked hard, were committed to the project, and liked, related to, or were seen to be interested in farmers. This had implications for project success and impact, since community leaders tend to be better at executing the wishes of higher authority than in articulating grass-roots feelings and concerns. Looking at leaders (politicians, church leaders and extension workers) as government representatives, there was a tendency for communities to listen to these leaders carefully and to obey what they were told. Highly placed people made more farmers join FFS projects (see Box 3). In the belief that government is supposed to be their provider, farmers tended to link everything to government, and feared to disobey what they perceived as government orders through not taking part in projects in their areas. Leaders' speeches/remarks made during any ceremony greatly contributed to the sense of legitimacy and acceptability of the projects among the communities.

---

[13] Every Ugandan national above 18 years is an eligible voter by constitution. To be able to vote, however, the person must be resident, registered and in possession of a voter's card. The Voter's cards are issued by the electoral commission at national level. Voting is by secret ballots.

[14] Councilors at the different levels form the electoral college of the executive (save the chairperson) at that level i.e. it is among the councilors that the vice chairperson, secretaries for the different sectors like works, production, education etc are selected.

> *Box 3: Involvement of "big" people in FFS inauguration and field days to "buy in" more farmers in FFS projects.*
>
> In launching FAO-IPPM and MAK-SPUH, MAK-IPM the minister of agriculture, DAO and local leaders (at LCV and LCIII level) visited and talked to farmers during inauguration and field days. This is a motivation to farmers when they relate with "big" people. Handing over a spray pump and ox-plough as tokens to two high performing IPM-FFSs in Kumi district by the honorable minister of agriculture (by then Hon. Kisamba Mugerwa) as the guest of honour during one open/field day organized by Makerere and NARO/DFID/COARD project not only stimulated the desire of more farmers to engage in FFS but also raised the status of some FFS members in their communities. During one of the field days in one of the FAO-IPPM FFS, the sub-county chief offered a heifer to the group after being impressed by the work of the group. This boosted collective work for the group's sustainability.

Villages, parishes and sub-counties, and districts are linked in a local council structure, led by an elected chairperson. The local administrative structure is mandated to mobilize and serve community interests. It is charged with the responsibility of over-seeing implementation of government programmes (in any field) at the grass roots. At village level, the leaders are close to the people (they are locally resident) and practically know where any named individual lives. These local leaders are also farmers too, born and resident in the villages they lead (these are requirements for a candidate) therefore truly part of the community, but with better than average resources (in terms of land, income, better housing, and literacy). This is perhaps one of the reasons why they are respected (or feared) in the community.

Wealth is power. Even when local leaders were not fully involved or interested in a venture, informing them was one way to acknowledge their power and influence, and to show respect for their status. Invitation of and attendance by leaders, especially from the higher political levels, such as members of parliament, district chair persons and sub-county leaders (as explained at the head of this section) attracted more farmers and in turn created the impression that very many farmers were interested or involved in the projects. The likely presence of the 'big shot' is announced prior to the function, creating a sense that attendance of the community is obligatory. Leaders at lower levels also stepped up their community mobilization activity with the knowledge their bosses were involved - levels of mobilization and community attendance seemed to be directly proportional to the status level of the leader invited to preside over the FFS, and the numbers of representatives from partner institutions like FAO, CIP, or even a funding agency at the international level, likely to attend. Mosse (2005) has described similar carnivals of legitimacy as an aspect of participatory development in India. Presence of a representative from the major development institutions implies the meeting or venture is blessed and therefore valuable. The community felt honoured when such people travelled from urban areas to the rural areas to meet and talk with them. Similarly, politicians and aid bureaucrats felt tangible political support, and were thus reassured that development was "working". Whether and what this has to do with discovery-based learning and resolution of hard choices involved in technology development we will later examine.

2.5 International interests in shaping and re-shaping FFS

The main concern in the present thesis is to understand how FFS projects are structured, organized, and implemented when viewed from the field perspective. To do so, we need to understand project inputs (at the top), outputs (at the local level) and the changes that go on as projects move down the hierarchy from donors to target communities. Although interests and perceptions at all levels contribute to the understanding and implementation of FFS projects, researchers at the international level greatly influence the overall objectives of such projects. This could be attributed to their role and skill in negotiating with donors as well as players at the national level. A paradox that these actors face is that in order to create a space for participation they need to open up an operational opportunities with both donors and government, but in order to do so they often have to frame proposals in terms of technology perceived to offer the most appropriate solutions to problems already "on the radar".

At international and national level, FAO staff (especially at national level) were key figures in influencing the agency "climate" towards FFS in almost all potential partner organizations in Uganda. A major way in which this worked was when key individuals working with FAO were called upon as resource persons in assessing and setting up FFS interventions. Consultation with senior FAO provided interested parties with general insights about the farmer's context. It would be explained that such information might not be sufficiently up to-date, specific, realistic or truly representative of the farmers' perspective. At this point the international advisers would typically call for the introduction of facilitators and others who had already taken part in successful FFS activities elsewhere to help build better understanding of what might be workable in the local context. This provided the means to convince sceptics that FFS projects were "do-able" in Ugandan conditions, but at the same time served to reproduce an international FFS "style".

The next step involved mobilization of national researchers for involvement in FFS. This introduced subtle changes. At national level, researchers took on FFS projects, with some cues from above (e.g. agencies such as CIP), as an opportunity to disseminate, evaluate or try out technologies (particularly improved varieties, such as the orange-fleshed sweet potato and associated agronomic practices (evidence is presented in later chapters). The addition of technology dissemination opportunities, and possibilities for evaluation by farmers, proved an incentive for scientists at national level to engage in FFS projects. But this tended (as we will later see) to give project and project objectives a different shape - an orientation towards dissemination and use of improved varieties, and agronomic practices, rather than discovery-based learning. It would have been hard to resist such additions, however, because they actually complemented IPM related integrated production technologies (IPPM and IPPHM) already envisaged as part of the Ugandan FFS. This created a situation in which the interests of researchers are perhaps best characterised as small projects within a bigger project. At local level, selection of areas (villages) and people (extension staff of facilitators and farmers) vital to project success, also influenced, or reshaped, implementation.

Once a hierarchy has been established - and FFS in Uganda is a hierarchy (Figure 3), even though its theme is "participation" - everybody further down will wish to play a role, to signify recognition of authority. Though left out in negotiations, farmers take on any project as a mark of respect to authority and in the hope it might just offer some resources that can be captured and redirected towards solution of their real problems. A central point to be explored in the further chapters of this thesis is that there are no guarantees that these local problems will in any way relate to the perceptions built into the programme design and rationale from "on high". The preliminary conclusion to emerge from this brief technographic sketch of FFS in Uganda is that if it works (a question for the rest of this thesis) it will be in spite of the institutional features it has acquired as a result of translation into Ugandan practice.

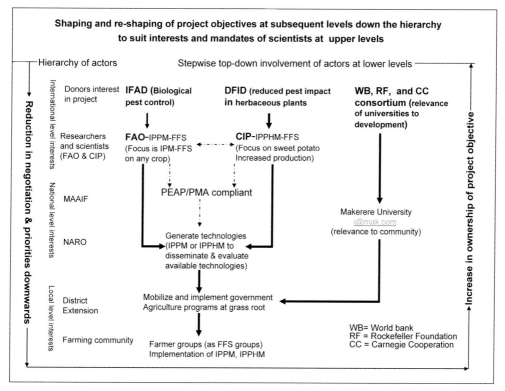

Figure 3: Shaping and re-shaping project/objectives to suit interests at different levels.

2.6 Concluding remarks

This chapter has attempted a technographic sketch of the main organizational features of FFS as a system for discovery-based learning about agro-technology in Uganda. Analysis of actors and their interactions indicates that there is a hierarchy of actors - from international,

national, to local levels. Identification of farmers' problems, and prescription of solutions by scientists, gives rise to a number of issues. One is that those who frame development projects are entangled in the thinking of existing institutions, including the agenda of existing scientific programmes. In working in hierarchical conditions, dictated by the "top-down" politics of international development aid, project planners maintain a hierarchical mindset which affects even supposedly "participatory" development. Powerful, well-connected international actors decide upon what is believed to work as the best technologies while colleagues at national and local levels with similar interests and methods of operation, cooperate in order to relay technologies to farmers. Instead of the two-way learning platform envisaged by populists when FFS was first conceived, "actually existing" FFS seems in danger of serving the interests of those researchers who continue to administer a conventional top-down linear technology dissemination approach (technology transfer), but by mobilizing local actors more effectively than hitherto. This contradicts the very basis of FFS as a methodology to gain farmers' participation and insights as an active force in technology development.

Under mandatory rules of system operation, and a conventional mindset, participation risks remaining theoretical window dressing (as subsequent chapters will substantiate). Participation in problem identification and prioritization does little more than create a stepwise (rather than linear) hierarchy for technology dissemination. Research institutions enthusiastically embrace FFS, but more as a conduit for transfer of their technology under the guise of participatory learning. FFS then is in risk of becoming a platform in which researchers promote the mandates and objectives of their institutions rather than actually address farmers' interests. Implementation, it is suggested, works mainly through local elites, questioning claims to inclusiveness. How much this risk is real will be documented below. Conceptually, FFS was meant to be a collaborative activity but where implementation is top-down farmers remain at the receiving end. Participation becomes a stepwise invitation from the top. As Cornwall (2004) puts it, getting a seat at the table does not necessarily mean having a voice. Farmers (at the local level) are very important in implementation of project activities, but their interests are rarely, if ever, taken up as a starting point for a project. At the extreme, participation risks becoming a means to mobilise a labour force for the achievement of goals set by others. This dubious possibility has been glimpsed via the present technographic survey, and it will be the business of subsequent chapters to assess whether or not the danger is real. These chapters will contribute to addressing a concern voiced by Cooke and Kothari (2001) that in reinforcing rather than challenging power relations, participatory development falls short of its own declared goal of empowerment. The key challenge in FFS, we will eventually conclude, is how effectively to involve farmers in problem identification and prioritization processes right from the point of project conception at the international and/or national levels. Penetrating the rhetoric of participation, understanding what goes on at the local level, and changing mindsets to reinforce belief in the poor as agents of their own technological empowerment are among prerequisites for a reformed system to be glimpsed once the local-level functioning of Ugandan FFS has been analysed in further detail.

CHAPTER THREE

New technological inputs and local farming activities: mobilizing actors and instruments for FFS

3.1 Introduction

As discussed in chapter one, technology is the intersection point for the realities in which FFS is embedded. Technology can be defined broadly as human instrumentality. As such it typically involves a nexus of human agents, tools, instruments and processes, and associated knowledge. By the term "new technological inputs" we refer to additions to the stock of instrumentalities and knowledge. These can emerge locally (from modifications of local practices) or from external sources (technology transfer). In this work we will focus on new or improved agro-technical instrumentalities, derived both from agricultural research and local sources, resulting in planting material and general agronomic practices different from traditional or established practices, and contributing to improved agricultural productivity. Through FFS, attempts have been made to transform and disseminate various agricultural technologies with the aim of improving upon yields, hence positively affecting food and income security among the Ugandan rural poor. The technologies range from improved crop varieties to better production and post harvest handling practices. To realize optimum benefit from the new technologies, farmers are mobilised by competent people who undergo a preparatory process to enable them to become effective in stimulating the discovery or adoption of new technological inputs. Under FFS, farmers also play a key role in shaping new technological inputs. FFS envisages that they, too, can contribute to the invention or adaptation of technologies to suit their local contexts. In FFS (as opposed to other kinds of participatory rural development) mobilisation and training are centred on new or improved technology. While we discussed how institutional interests or objectives shaped FFS projects in chapter two, the issue in chapter three is how technology is central in defining activities on the ground. Following the actor-network theory (Law, 1992) technology becomes an actor. FFS can be seen as an exercise in inducing changes and extensions of actor networks around technology for the benefit of the rural poor.

It is important to specify the types of technologies or interventions covered under FFS in Uganda, the rationale for choice of specific technological intervention points, and the mobilisation process involving facilitators and farmers, in order to arrive at a clear analysis of how the nexus of technology and society (the socio-technological ensemble) in rural Uganda can be advanced. Analysis of this information gives insights into how new technological inputs link to local and "outsider" interests. This work is not a formal evaluation of new technological inputs, or how well they work, but a technographically (i.e. descriptively) oriented account of preliminary activities and processes of interaction triggered via FFS implementation. An overall conclusion will eventually be reached that FFS does not easily translate into a process where farmers take an active role in choosing and developing technologies to address their local situation, interests and needs. What will be demonstrated is that the institutional resources upon which Ugandan FFS has to rely are already configured around assumptions of technology transfer (ToT). In the ToT approach technology is seen not as instrumentality but as commodity (a package, a product, a machine). The logic of an on-going market revolution (biased towards already manufactured technologies) all too readily overwhelms the basic FFS notion of discovery-based learning. FFS is adapted to become an opportunity to disseminate what is already in stock rather than a chance to develop technologies that suit local realities. And yet the discovery-based learning approach is apparently needed. Those in charge of re-directing agricultural service delivery in Uganda (under the NAADS programme) are reportedly surprised to find out that supplies of "on the shelf" technology ready to roll out to small-scale Ugandan farmers are more limited than they had imagined (Bukenya, 2007, forthcoming). A range of issues concerning needs for new technological inputs and mobilization processes need to be rethought if FFS is to achieve its overall objective of effecting a junction between the best available innovations and local adaptive inventiveness.

3.2 Technology interventions covered by FFS in Uganda

The main technological entry point in FFS has been improved crop varieties with higher yields and resistance to pests. From examining the range of interventions covered by FFS projects in Uganda (Table 1), it will be seen straight away that all the technologies were already in existence in some form or other. There are no cases where FFS intervention in Uganda has resulted in setting up new technological intervention programmes to address local conditions and realities markedly different from the conditions under which the "on the shelf" technologies have been conceived. Adaptation and *in situ* inventions are more difficult and time consuming. Everyone (farmers included) would prefer to make use of what is available on the shelf, and this fits the institutional top-down biases of a system in which functionaries have not been trained in participation or participatory technology development. Agricultural professionals as yet lack adequate skills to engage in participatory processes, as is well analysed in a recent study on competence building for Makerere University lecturers designed to improve training of development professionals (Kibwika, 2006). In addition to institutional background, lack of time and resources makes in-situ technology development difficult. Donor perspectives and project planning cycles (typically 3-5 years) are often too short to show real results from *in situ* technology development starting from scratch ("scratch" often being a good definition of the problem). In truth, problem definition alone might take several years. This, however, is not to say that FFS is too labour or skill intensive in the demands it makes on institutional creativity. Farmers do invent. Cases are known where technologies have spread without official intervention, sometimes as a result of having been "borrowed" or "stolen" from experiments that never led to any official release (cf. Richards, 1985; Jusu, 1999). Clarity and commitment concerning objectives and procedures for adapting a given technology to meet specific local needs at the outset, and fuller analysis of the local context, to understand the real constraints to active engagement of concerned farmers, are important issues that if addressed could make better use of limited time and resources in catalysing in-situ technology development processes.

Table 1:The range of interventions covered by FFS in Uganda.

FFS project	Study or entry Crops	Technologies disseminated to meet the objective of the project
IPPM - Integrated Production and Pest Management	Cotton, vegetables (cabbages, kale, onions, tomatoes), cassava, groundnuts, maize	Improved and new varieties; recommended seed rate and spacing; timely planting, weeding and harvesting; use of manure
IPPHM - Integrated Production and Post Harvest Management	Sweet potatoes	Orange fleshed varieties; planting method (use of ridges); vine length (30cm); isolation distance between old and new fields (wider or intercropped with a cereal); Rapid Vine Multiplication; sweetpotato processing (value addition) - marketing
ISPI - Integrated Soil Productivity Improvement	Groundnuts and maize	Improved and new varieties (Serenut-II groundnuts, Longe 5 maize varieties); use of Mucuna, Carnivalia, lablab, Tithonia & *Lantana camara* as organic fertilizers; Farm Yard Manure (FYM) as organic manure; Single Super Phosphate (SSP), Di-ammonium phosphate (DAP) and Urea as inorganic fertilizers
IPM - Integrated Pest Management	Cowpea and groundnuts	Improved varieties (Igola-I, Serenut I-R and Serenut II groundnuts varieties and MU-93 cowpea variety); recommended spacing and seed rate; three sprays (at budding/vegetative, flowering and podding stages)
SPUH - Safe Pesticide Use and Handling	Vegetable (especially tomatoes)	Improved varieties (Heinz and Manglobe tomatoes); recommended spacing; site selection; staking; mulching; pruning; protective gear

FFS in Uganda does not have to condemn itself to trip over the typical obstacle to ToT - the expensive or impossible task of fitting the environment (including the social environment) to an existing commodity or invention.

The real purpose of FFS is to support or trigger the spontaneous spread of new technological inputs, whether from without or within. Where a technology is both suited to local conditions and relevant to the needs of farmers they will often fight to obtain it, and to solve any subsequent adaptive problems. Classic African instances remain cocoa and cassava (Richards, 1985). In the first instance these two South American crops spread in Africa without any official approval or support, and (in the case of cassava) against the wishes of some colonial regimes. Because these crops fitted local needs farmers were assiduous and effective in adapting local institutions of land and labour, processing techniques and marketing mechanisms. But both these crops spread in the early colonial period when there were spaces (by default) for farmer mobilization around technological issues, e.g. the spread of cassava processing technology along the "Ijesha Road" - an African missionary network in the Yorubaland Protectorate not yet fully controlled by the state (Richards, 1985). Today, the state is more pervasive, and farmer innovation has declined.

What FFS seeks to do is to create a mechanism through which farmers re-convene to engage in technological experimentation. In order to develop technologies that are useful in local contexts, we should think more in terms of "pump priming" or catalysis of local/poor people's innovations, than hands-on educational interventions promoted to support ToT. What FFS seeks to do, in theory at least, is to find the points of catalysis where groups of farmers combine to continuously modify interventions to fit their specific needs and contexts. This need not preclude "on the shelf" technology, but whatever is drawn down from the shelf should be understood as prototypes, and not already made or finished products. "Pre-cooked" technologies and prototypes (as will be shown) support only rather limited (and unenthusiastic) actor-network interactions among farmers. The data below rather suggest that "pre-cooked" technologies commonly used in Ugandan FFS limit room for interactions and agency at the lower levels, instead of stimulating it. Prototype technologies (it can be argued) will create more room for interaction as they are continuously developed, and modified into a product that suits the local context. Unfortunately, in the Ugandan application of FFS we could find only two perspectives, both top-down rather than discovery-based: the crop targeted intervention and the integrated crop package. We will now examine in detail how these kinds of interventions actually operated.

3.2.1 The crop level perspective

As already shown in chapter 2, researchers were an important constituency in bringing FFS to the field in Uganda. Researchers made key decisions about which crops to work with, as study crops. Each FFS project was written in such a way (perhaps to attract donors) that focus on important crops in specific areas formed the basis of the project document. This in turn implied certain implementation sites favourable to such an agenda. The crops were mainly those upon which the scientists in the project had more expertise. Typical scenarios included ones in which the crop in question had been researched to some extent but more work was needed, or research results were ready to be tested at the on-farm level, or results were ready for dissemination to the wider farming community. Examples are sweet potato, potato under CIP/NARO, and tomatoes and legumes (cowpeas and groundnuts) from Makerere University. Given this "lead" then prospective project sites were chosen by implication - i.e. where the crop was an important food, cash or both a food and cash crop. For example cotton and sweet potatoes selected for

Soroti, potatoes for Kabale, groundnuts for Iganga, cowpea for Pallisa and Kumi, soil fertility for Busia and Tororo, and vegetables (tomatoes and cabbage), therefore pesticide use, for Mukono and Kiboga. In no case, could we find evidence that FFS was seen as an entitlement of rural communities, and communities then asked to specify the topic of choice. Where community choice elements came into problem definition this was at the very lowest level and last moment. For example, the Mukono and Kiboga groups had a choice of whether to take on tomatoes or cabbage, although even here we discovered that this was mainly influenced by the facilitators. For example in one FFS, choosing tomatoes, we discovered the facilitator was more informed about tomatoes, given that his M.Sc. degree was on this crop.

In Tanzania and Kenya under the FAO-IPPM study the entry crops were bananas and tomatoes respectively. It can be wondered whether the choices would be the same if farmers were given an opportunity to determine their study crop. The issue deserves some thought. Choosing the most important crops in the different communities in which FFS projects operated probably was a strategy to engage as many farmers and other players as possible, under the assumption that the main crop will touch the majority. The underlying rationale seems to be "choose the project sites that include the main producing areas (in the district or village) of a chosen project entry crop, or the village most hit by the topic of concern (disease, pest, soil fertility etc) in relation to the crop". A crop being important in a given area assumes that farmers are homogenous, yet in reality there are many differences across individual farms, villages and communities, based on various factors including land access, religion and ethnicity. The single most important factor within communities is often socio-economic variation. A region may have a speciality - cotton or cocoa, say - but these may be crops from which the poorest farmers are systematically excluded (they may not be able to afford the inputs, or they might not have the requisite land rights - women in many parts of African cannot own trees, for example). Even where the crop is widely grown its regional status does not necessarily make it important to every one in the same area in the same way. Again, it is not hard to see that the popular view might then be at variance with the views of scientist/researchers about entry points for FFS.

Changes in social and economic conditions facing farmers contribute to variability in crop importance over time. Example in this context is the cotton crop that was chosen by FAO-IPPM project in Soroti and Busia. The crops chosen in Kenya (vegetables) and Tanzania (bananas) remained of interest to farmers before, during and after the project interventions. But the story of cotton in Uganda was different. Many farmers did not engage much with it even when the project started, and the decline of interest in growing cotton intensified during the project period. IPM is a technical concept and choice of a crop like cotton with a wider range of pests offers a good opportunity for farmers and facilitators to get a clearer understanding about the basic principles of pest identification and management in (cotton) crop production, but this is of little general significance when the crop itself is of diminishing interest due to decline in market prices and environmental instability. Governments, however, see the matter differently. Cotton is a valued revenue earning (i.e. taxable) crop, so there is a tendency for governments to welcome FFS applications on cotton as a way of reducing costs (e.g. for pesticide imports) and thus of re-booting an important export trade. But this would imply FFS being used to induce farmers to grow a crop in the face of adverse market signals - hardly consistent with the overall aims of the reform programme in Uganda.

Failure to analyze the targeted community in relation to the proposed technology leads to waste of resources (time, land, money, energy) of all actors involved. But do researchers and other actors feel the waste as much as the farmers do? The idea behind IPPM in cotton was to minimize pesticide use and encourage organically produced cotton. There was no real local motivation for this venture because the organic alternative FFS was seeking to stimulate was very demanding in terms of time and labour inputs. These labour inputs were simply not feasible when viewed from the perspective of farmers with multiple responsibilities, inadequate quantities of organic residues, many fields (often distant from home), and domestic needs to attend to. By the late 1990's, when the project was initiated, most farmers in eastern Uganda had given up cotton production due to bad experiences associated with the government-regulated marketing system. Farmers used to take their cotton to a collective buying/selling centre, i.e. a cotton cooperative society storage site in their area. Despite the high production costs incurred in cultivating cotton (planting, weeding, spraying, harvesting and ferrying crops to the cooperative society) prices declined to discouraging levels. These low prices were also paid very late, and some unlucky farmers had no payment for their cotton at all. The interest of farmers in cotton production faded. The response of farmers, even those in FFS groups, to cotton production was negligible. To most farmers, cotton was no longer worthwhile. In this context, working on the marketing of cotton rather than pest control might have made better sense of farmer concerns.

Not to lose farmers' interest in the project, other food crops were taken on. These included cassava (management of the mosaic problem) and vegetables, groundnuts, maize and sorghum. This flexibility helped capture farmers' interest to some extent, so that there was some willingness to engage with project technology, given the importance of the new crops included as both food and cash crops. This, of course, shows that not all FFS projects in Uganda were insensitive to farmers' concerns, but it would have been better to make these concerns as priority from the outset. Other projects (CIP and Makerere) stuck to the same crop/technology throughout the project time, where FAO shifted crops. This might suggest CIP and MAK chose technologies with which farmers were more closely engaged.

Although farmers under FAO-FFS did, in the event, exercise some influence over project choices, we need to examine carefully the context in which that apparent influence was exercised. The FFS secretariat in the districts made project adjustments based on how the extension facilitators perceived the situation. It was argued that facilitators were more exposed to and therefore more informed about the overall district situation than farmers. One DAO stated that "...the extension staff facilitators are more knowledgeable than farmers and are in the best position to guide the farmers on the most appropriate enterprises that will help them increase food and generate income in their home...", during one of the meetings with the FFS facilitators of Busia. The vegetable enterprise, chosen by the FFS secretariat, was perceived as a lucrative crop of help to households in Busia in generating income. Cabbage, Kale (*Sukuma wiki*), onions and tomatoes constituted the crops chosen for the vegetable enterprise. *Sukuma wiki* is a staple leafy vegetable in Kenya, introduced into Uganda through the FFS. Commercial objectives were part of the rationale, since Busia is on the Kenyan border. To improve upon knowledge of farmer-facilitators in vegetable production, a training session was organized and the resource persons were extension facilitators from Kenya. Farmers in the FFS groups did not choose the vegetable crops, although they cultivated them on their farms even before FFS initiative in the area, on a very small scale. Again, it seems clear that there was little scope for farmers to choose what they were offered.

Once this choice for vegetables had been made, certain logic clicked into place. Prioritizing vegetable production (by the district FFS secretariat) offered FFS groups an opportunity to take up such enterprise as a commercial activity. But here lay a problem. Commercialization of vegetables calls for farmers to be interested in and to value a given crop. Busia has soils of relatively low fertility not well suited to vegetable production. Leafy vegetables provide a cheap source of food relish[15]. Food relish is one of the limiting factors that people in rural areas face especially during the dry season. The staple relish in Busia is fish, which is expensive, and not affordable by many people, whether from the farming or non-farming community. Leafy vegetable like kale would be a handier, cheaper substitute, but farmers were reluctant to take this up mainly because it was not liked by many. Besides, the poor soil fertility levels did not favour good performance by the crop. This would mean using fertilizers to boost upon soil fertility.

In addition to expensive seed and unreliable agro-stockists, vegetables are very susceptible to pests and diseases, and need fertile and moist seedbeds, something farmers in this context were not able to supply. "...If you take up production of onions, you will generate more money and will be more of my friend...", mentioned one of the facilitators as she passed through a market where onions were being sold in the presence of one of the FFS group members in Busia- Uganda. This statement, on one hand, shows that the facilitator actually cared about FFS farmers and desired to see food and income security prevail in the community. On the other hand, it also makes clear the facilitator's presumption to influence the choice of enterprise perceived to be beneficial to farmers. In this case, farmers are still looked on as people who cannot make constructive decisions for themselves.

3.2.2 Package level: the integrated approach perspective

In current research and development discourse integrated approaches are often advocated, for efficiency and effectiveness. The same approach can be applied to research institutes, as demonstrated by de Janvry and Kassam (2004). In attempts to improve upon agricultural production, integration, and multi-disciplinarity is the trend. No single component can exclusively solve prevailing farming problems in isolation, since farmer's situations in themselves are complex and interrelated. The integrated approach is rich in complementary purposes, a reason why several FFS projects use the integrated approach. An integrated approach views a

---

[15] In Uganda, dinner and/or lunch food basically must have at least two things: a carbohydrate in solid form (derived from bananas, tubers, and cereals such as maize (for posho), millet, sorghum, and rice) that forms the dry food, and an accompaniment based on legume, animal flesh (meat, fish, chicken) and vegetable (leafy) stew as a relish to make the whole meal tasty and enjoyable.

systems perspective as one way to deal with problems. In the systems perspective interactions of elements are important, and more than a sum of the parts. The systems perspective only exposes farmers to a variety of alternatives that can be used to improve crop production and protection practices but also reveals some more effective ways of approaching improved crop production through combinations of specific components that farmers might have earlier either ignored or not thought about. The systems approach can thus be thought of as a scatter gun, aiming at multiple targets. As to whether these diverse targets fit farmers' contexts or their own conceptualisations of the problem is a different issue.

Components of the integrated approach included improved seed varieties, site selection and preparation, planting method and time, pest and weed management, harvesting and storage, soil fertility management, and minimal pesticide use. These apply across all projects. A real systems approach (using some kind of input-output modelling approach) was not visible, perhaps because of the researchers and technologies involved. As a result, the degree of emphasis given to specific components (soil, pests, and pesticide use) compared to other(s) seemed to depend rather heavily on the areas of expertise/focus of the "mix" of researchers in a given project. Information sharing between projects created room for adjustments of packages - mainly in the form of "add ons". For example, CIP solicited information from FAO-IPPM-FFS facilitators from ICM, IPPM and IPPHM, to cater for processing or value addition in the case of sweet potato tubers. Possibly ICM showed the way to go, since it had a broader coverage, embracing soils, crop production activities, pests, and post-harvest handling. Each of these is a system in itself, though affected by all the rest. In spite of integrated designs focusing on soil, production, and post harvest handling, safe pesticide handling and insect pest management technologies dominated in all FFS projects, mainly due to strong influence from FAO and its own background in IPM.

Sharing of information - i.e. horizontal learning - by project implementers and former FFS cadres at national level - created room for adjustments to cater for issues like marketing that were of interest to the community. These adjustments applied to different crops and areas. Marketing, registered as a need in cotton production during FAO-IPPM-FFS, was picked up on and incorporated in CIP-IPPHM on sweet potato production. In sweet potato production, under the CIP-IPPHM project, the focus was more on processing to increase commercial value of sweet potato. Sweet potato production in the study sites was mainly a subsistence activity and farmers did not show as much interest as was expected. There was no ready market. It is more feasible to produce after securing markets than to search for markets while sitting at home with a rotting pile of perishables. Presence of mills as potential markets did not necessarily mean presence of actual markets, since farmers failed to locate transport or buying agents and often got stuck with potato chips in their courtyards.

The integrated approach in Ugandan FFS was typically aggregative, i.e. it combined results from different and often independently conducted research studies on the same crop, covering breeding for desired characteristics, recommended agronomic practices, entomological studies focusing on specific (important) insects, pathological studies with respect to specific diseases perceived to be of high importance, and analysis of soil fertility requirements among others. Promising outcomes for insertion in integrated packages were then tried out piece-meal on-station, and promising technologies released for on-farm trials, evaluation and dissemination. On-farm experimentation was often with individual contact farmers, or with a few FFS groups. During on-farm trials and evaluations, practices externally (including farmer practices) were compared with farmers' local practices. In this way, farmers were exposed to a variety of technologies from which to make informed decisions suited to their individual contexts. But prospects for farmers taking up whole packages as introduced were often minimal, generally on account of highly varied social and economic situations. Even so, the integrated approach might be judged better than the crop-focused approach since it offered farmers a range of components or combination of components from which to choose.

The main problem with the integrated approach remains the commodity or package view of technology as a "thing", rather than seeing technology as a set of instrumentalities for achieving social or economic ends. This componental view of technology - even when arrangements are made to combine and re-combine components - may not necessarily provide solutions to actual problems faced by farmers. The elements may be useful, but they have to come together in the field, and serve some specific local need. Put another way, the elements of a technological solution require "hardware" elements to be coupled together within a specific social project. Innovation is as much a matter of providing space for experimentation around these social elements as it is a question of providing "hardware" to begin with.

This is why a second phase FAO-FFS project intends to integrate social and economic aspects in the FFS. A conversation with one of the project implementers suggests that the new programme will put more emphasis on farming as a business, savings and credit, collective marketing, revolving funds and capacity building for FFS alumni networks. How far this represents a real change of heart is open to some doubt. The interviewee made clear that the traditional FFS process was to be maintained on any crop chosen by farmers. But still, it was conceded that there was need to understand the way farmers related with whichever crop they chose, because this was a way of highlighting the relevance of the concept of farming-as-a-business in that context (true entrepreneurs make their own choices!). This process of integrating technology (crop and package) with social "capital" (e.g. emphasis on innovation and marketing chains) may creates more chances for a sustainable innovation process, given that the two sets of elements are linked and influence each other. Where the element of doubt, arises, however, is that this represents no real break with the "instructional" model. Poor people are deemed to be poor through their own lack of capacity. If this lack of capacity is more lack of business skills than lack of the right kit, then FFS should teach (preach?) business skills and not agro-technical competence. This misses out two issues FFS was meant to address - gaps in agro-technological knowledge, especially in addressing difficult local conditions (Richards, 1985) and, second, empowerment and justice (the idea that poor people may be poor because they do not control their own resources or destiny).

3.3 Preparation of FFS Facilitators

Technologies formed the basis upon which facilitators were trained, in terms of both content or curriculum and method of training. This section will trace out the process through which facilitators were identified, and how they were prepared or trained for the task. Reducing facilitation in FFS to the task of teaching farmers about how best to use pre-manufactured technologies from researchers tended to reproduce the conventional way of doing extension. In re-orienting extension through FFS, facilitation needed to be rethought. It should have been geared towards building competence of extension workers in understanding or analyzing the local context in which the technologies were to be introduced (cf. Kibwika, 2006). If technology is instrumentality and not equipment then facilitators need to be in a position to identify and work with social relations of production in order to offer more interactive opportunities. Technical skills alone do not prepare facilitators to choose the most appropriate ways of applying the technical skills.

3.3.1 Identification and selection of facilitators

Competence in facilitation is enhanced as much by interest and will to succeed as by possession of relevant training geared especially to inculcation of the social skills required to promote interactions between actors. There is need for criteria with which to select prospective facilitators. Quality of facilitators will heavily depend on the selection criteria used, and the training content and approach. Personality, interest, empathy, commitment, and willingness to work with people are important aspects in facilitation. This clearly brings out the need to understand the people and context one is to work with. Braakman and Edwards (2002) confirm the importance of interest in the audience (farmers and their context, in this case), listening carefully to what others say, and willingness to change oneself, are very important in building facilitation skills. Working in partnership to minimize parallel programs and integrate FFS activities into the district production sector, FFS projects left the responsibility of facilitator selection to the districts. During the selection of prospective facilitators, no clear criteria seem to have been followed.

DECs selected extension workers while extension workers (then facilitators) selected farmer facilitators, which had implications on sites to be chosen to participate in FFS projects. For the case of a CIP-IPPHM sister project in Bukoba (Tanzania), at the time of group formation, farmers were given the opportunity to select whom they wanted as their facilitator. The program assistant gave them some criteria to guide the selection. Often, they selected their village extension workers (VEW). This probably was due to the fact that they met the VEW more frequently and knew them. It seems unlikely that they had much knowledge of other potential candidates. The project assistant then requested the selected VEW formally from the DAO, who sometimes substituted the selected VEW with another VEW, depending on needs within the larger system.

Some projects like MAK-IPM and MAK-SPUH had some influence over selection of extension workers to work with as prospective facilitators. Priority was given to recently graduated students who were once under project staff supervision while at university. Subsequent FFS projects (like CIP-IPPHM and A2N-ISPI in the same districts) took on facilitators formerly trained by FAO: these facilitators were referred to as core/national staff, and could serve any FFS project as facilitators anywhere. Although taking on already trained facilitators was more convenient to subsequent projects and built individual competences in facilitation, there was a corresponding (negative) of serial project abuse by some individuals working minimal changes, from posting to posting. One would expect the core facilitators to understand the farmers' local situation, and to be able either to advise researchers on what was happening on the ground, but not all had the skills to work in this way. Nor were they at times very keen to perceive what farmers do as appropriate to the local situation, since they had imbibed what research and extension preaches - that recommended practices are inherently superior to farmer practices. Some of these perceptions towards farmers are reflected in the way they relate with farmers (Box 4).

Box 4: Relationship between facilitators with farmers.

The relations between facilitators and farmers varied. There are facilitators who relate well with the farmers and are committed to promote interactions around local technologies. But there were others who took improved/new technologies as an excuse to abuse farmers and the projects in a number of ways. Such facilitators had different priorities, and did not meet the FFS groups as frequently as agreed, but nevertheless demanded full pay! In situations where farmers hesitated to pay them, such facilitators uttered threata. "I will remove this project and take it to another group if you are not cooperative". "If you do not give me the money, you will see". The farmers yielded because they feared they would lose opportunities. One facilitator even demanded to be paid in dollars! He told the FFS group under him that the Ugandan currency (shillings) had no value against the dollar unless adjusted - upwards. Farmers did not know anything about dollar-shilling rates and were supposed to pay the facilitator at a fixed rate in local currency. In this case, this facilitator used the promise of project technology to lever money.

Under SP PPHM, one facilitator was given money for 2 FFS groups that he claimed were ready to work and he was the facilitator. After receiving the money, he used it for his own personal expenses, and the farmers lost out. The facilitator kept dodging the programme assistant with lame excuses. Other facilitators used English when defining some concepts to the farmers. I witnessed this during two sessions in one ISPI FFS in Busia when the facilitator was defining soil. "Soil is a natural resource..." he told farmers (in English) who had gathered for one of the introductory sessions. Farmers just looked at him. In a typical rural community where a majority (more than 80%) did not go to school, it makes no sense to use English, yet the facilitator could speak the local language. His utterance was clearly intended to impress rather than to communicate information. The interesting thing is that the three extension staff facilitators above were all once FAO-IPPM trained. It might be better to take and train truly interested, responsible and self motivated people as facilitators other than re-cycling people just because they are designated core facilitators under the FAO-IPPM-FFS project. The wrong kind of people look at projects as opportunities solely to make extra money but not to increase relevance of technologies to a community. It seems implausible to expect facilitators with such attitudes ever to facilitate even ToT let alone anything approximating to discovery-based learning.

3.3.2 Perceived selection criteria for extension staff facilitators

Discussions with facilitators, especially under FAO-IPPM, revealed that they actually had little idea as to why they were selected by their DAO to become FFS facilitators. They just received invitations to go for training and could not refuse. However, they thought of various reasons for their selection, ranging from commitment to work, status, specialism or ownership of a means of transport (e.g. coordinator of extension activities, crops officer, livestock production officers, trainer at a farmers' training centre (FTC), possession of a motor bike, and participation in the program in its initial stages). There was a feeling that choice of either committed or less committed extension staff to work with some FFS projects depended on how well the DAO related with the program assistant, and the DAO's perception towards the project (the more positive the perception of the DAO about project or technology it espoused the more committed the extension workers serving as facilitators would tend to be. However, this investigation also encountered a tendency for the more active, responsible and hardworking extension workers to be retained in government projects and programmes. Good work in extension means an ability to convince many farmers to adopt[16] improved technologies and recommended practices irrespective of relevance in the local context. Priority is given to government projects, given that the public extension system is more accountable to government, which employs them anyway. Some DAO's probably wanted to retain those field officers capable of delivering on targets when these were demanded by government (perhaps to convince donors). This probably explains some of the differences in facilitator commitment to FFS projects across the different districts.

Although education level of extension staff facilitators did not vary a lot, given that majority were diploma holders (Table 2), it was observed that graduates tended to be more responsible and accountable. Projects (A2N-ISPI, MAK-IPM and MAK-SPUH) using graduates bore witness to this. The few graduates that were given chance to co-facilitate with the trained diploma holders under FAO-IPPM in Soroti district, even though lacking prior FFS facilitator training, were willing to learn how to facilitate, proved friendly, empathetic, and supportive, and made regular reports. "They are responsible and with computer literacy will make it easier to have regular and well written reports" was a remark made by one program assistant.

By the time the FAO project was started, the majority of the agriculture field extension workers (FEWs) at sub-county level were diploma holders. There was a recruitment ban on agricultural extension workers at the time (1999). All trained extension staff facilitators were diploma holders. Degree holders joined the agricultural extension service when the plan for modernization of agriculture (PMA) was activated around the year 2000, when degree holders

---

[16] ToT is characteristic of conventional agricultural extension and research, where farmers are expected to take up technologies with all instructions as recommended by research. However, farmers' socio-economic situations never allow full adoption. Farmers take up technologies with modifications or alterations in recommended practices. There is therefore more of an adaptation than adoption because technologies are never 'finished goods'; farmers continuously modify them to suit prevailing local contexts. In participatory oriented programs adaptation is central, since it is through these interactions that farmers think of about appropriate, compatible and more feasible ways of efficiently using technology.

Table 2: Qualifications of extension staff facilitators in Uganda.

Case/project	Total number of extension staff facilitators trained	% composition by academic qualifications	
		Diploma holders	Degree holders
FAO-IPPM	27	100	-
CIP-IPPHM**	2	100	-
A2N-ISPI***	15	87	13
MAK-IPM*	14	72	28
MAK-SPUH	15	73	27

* Under IPM, Makerere trained more extension workers from other districts during refresher courses
** The two in Soroti and *** six of 15 (in Busia) are national/core facilitators trained under FAO-IPPM

were deployed at all sub-counties following the PEAP policy aimed at commercializing the predominantly subsistence agricultural sector. Transition from subsistence to commercial agricultural production required knowledge and skills on the part of the extension staff covering facilitation of participatory approaches for effective use of appropriate technologies. This explains why the new FFS projects in Busia and Soroti gave priority to extension staff with prior experience in FFS, who happened to be diploma holders trained under the FAO pioneer project. Note that being referred to as "extension worker" should not be mistaken for the field workers being actually trained in agricultural extension work. Often researchers assume that extension workers are analytical and therefore understand farmers' contexts, a reason why they rely on them for dissemination. They are/were actually degree holders from a range of science faculties like agriculture, forestry, fisheries and veterinary medicine, with mainly atechnical orientation, and limited training or exposure to dealing with complex social situations in the field.

3.3.3 Criteria used for selecting farmer facilitators

Prospective farmer facilitators on the other hand were mainly selected by extension staff facilitators whose criteria seem academic rather than pactical. Emphasis was laid on incorporating 'bright' farmers (this was assessed as those who participated a lot in discussions and group activities, and who were literate and speak English). Attempt to take up technologies as taught in FFS was another criterion, given that these farmers would need to serve as models or good examples to the community. Most of these people turned out to be leaders of different groups facilitated at some stage or other by this particular extension staff facilitator. Several were local leaders of the village, and in some cases they were farmers who had hosted experiments in prior projects. Such people were more confident and many times more well off when compared to the rest of the farmers in the same group. Thus they had different perceptions of what farm problems mattered to ordinary farmers. Few "covert" mobiliers (those who operate in the "backstage" environment of the African village, cf. Murphy 1990) were incorporated.

The criteria are already skewed towards male farmers with prior FFS experience (Table 3). Males, though fewer in FFS groups, were more literate and took on most group leadership roles. In addition training periods necessitating travelling from home and staying off-farm for some days were difficult for females, given their major role in actually carrying out a majority of farming tasks (working in garden/fields on a daily basis, and preparing food for the family) not to mention other household responsibilities (e.g. caring for children, the sick and old). Volunteering to attend most FFS sessions and putting in practice what was taught in FFS also favoured the males, given that they had more "free time" and more resources (especially land and capital to acquire the improved technologies).

The criteria seemed to make an assumption that being a leader (a social position) automatically implied being sociable, with good communication skills, interested in FFS facilitation and having a better understanding of the situation on the ground. Leadership (in the African village, as anywhere) is about the exercise of power, not a matter of presentational skills. Village leaders are many times wealthier and do not engage in local practices on the same terms as the majority. Not only did their perceptions of problems differ from the rest, they also

Table 3: Characteristics of farmer facilitators under ISPI-FFS, trained in March 2004.

Identity of farmer (Only initials used)	Sex	Age	Highest Education	Position in former FFS Group
J	M	43	O' level*	Secretary
S	M	37	Primary	Mobilizer
P	M	36	O' level	Chair person
D	M	25	O' level	Vice Chair person
J (2)	M	45	O' level	Member
V	M	37	Primary	Secretary
J	M	49	O' level	Chair person
B	F	36	Primary	Chair person
J (3)	M	30	A' level**	Secretary
B	M	39	Primary	Member
C	M	45	Primary	Member
M	F	42	O' level	Treasurer
D	M	40	O' level	Member
J	M	36	O' level	Chairman

*'O' level means ordinary level in secondary school and it covers senior one to four

** 'A' level means advanced level in secondary school and it covers senior five and six

sometimes lacked basic facilitation skills. A facilitator can be a leader but the reverse is not necessarily true, especially where high office depends on age or family position.

Fellow farmers in the FFS group also contributed to selecting a prospective farmer facilitator. Under CIP-NARO-IDM, FFS alumni farmers volunteered to become farmer facilitators. This self-selection was mainly to help others improve their potato crop yield and thus help protect fields from being invaded by people who did not have food due to failure to control wilt disease. FFS alumni fields looked and performed better than other fields. Fear of losing the crop to thieves created a disquiet that impelled some farmers voluntarily to train other potato farmer groups in the area. Their aim was to protect their own harvest by ensuring the disease-resistant technology spread as fast as possible (an instance of where equipping the FFS process with social analytical capacity would have paid dividends, since it was clear that these farmers were proactive innovation diffusers capable of relieving the project of a burden to disseminate!). Other cases were different, however. Under A2N-ISPI and CIP-PPHM, farmers in the FFS agreed on whom they wanted to select as a facilitator. In some cases this created conflict. The preferences of farmers and facilitator came into conflict (Box 5).

Although farmers participated in selecting a farmer facilitator, it was not participatory for the farmer group to be facilitated by this farmer. The farmers selected did not express interest in being facilitators but found themselves selected because of the criteria used. It would be better to have identified and set up groups first, and before asking them to select who they wish to be their facilitator. It is difficult for farmers coming together for the first time to select a person. The difficulty eases after they have been interacting as a group for some time. The next step is that the selected farmers observe FFS sessions with a nearby FFS group, before proceeding for further training. We can now pose some questions about the feasibility of this next stage.

Box 5: Conflicting interests in choosing prospective farmer- facilitators.

The chairperson (also the local village leader) of Okunguro FFS under CIP-PPHM preferred someone different from the person the rest of the group wanted. The chair chose one of his wives. However, when he learnt that she would be the only female trainee, he retracted. "How can I send my wife among men? Will I know what they will be doing to her?" he explained, when asked why he gave up the ideas of his wife becoming a farmer facilitator. In another situation under A2N-ISPI, one FFS group chose a member who had a hearing impairment (one needed to shout for him to hear). They may have reasoned it would be harder to hide things if the project had to communicate with a deaf man, in the hearing of all. The facilitator (perhaps unsurprisingly) preferred someone who could hear well. Farmers' choice however was respected. The problem came in training. No allowance was made for the man's handicap. All trainees were treated equally. It is not clear how well he coped.

3.3.4 Training the facilitators

How a term is used or understood influences the way people behave in relation to it. During FFS training, every one who took part as a trainer of facilitators was referred to as a "facilitator", both by trainees and by organizers of the training. This sends a misleading signal (as misleading as referring to a beginning pre-medical student as "doctor"). Facilitation implies an ability to understand the local context and guide group choices in appropriate ways towards achievement of desired objectives or outcomes at local level. In the FFS context, the desired objective was to develop or increase the usefulness of technologies in enhancing agricultural productivity. Guiding interactions directed for specific or desired outcome requires facilitation capacities. These capacities (Braakman and Edwards, 2002) mainly have to do with understanding the dynamics and meaning of the way people interact with the crop/technology in question. Understanding your audience helps in choosing the most appropriate way to deal with it so as to enhance social space for desired interaction. And yet analytical skills needed for facilitators to understand practices in the local farming system were not part of the training content. Training of FFS facilitators focused on technical skills (e.g. how to use improved varieties, identify pests and injury or damage resulting from pest attack in an IPM context, and how to use better or recommended spacing) which prepared trainees more for activities connected with the conventional technology delivery model than for competence in analyzing their audience and its problems and potentialities. This was reflected in the content and approach of the exercises used for facilitator preparation.

Content of the training
The training curriculum and content, as examined in this study, emphasised technical skills concerning the different technology packages but not how best to pass on these skills to farmers. For instance under IPM content focussed on use of improved or new (groundnut and cowpea) varieties, recommended spacing for specific varieties, the recommended spray pattern and identification of the different insects (pests). Designing and laying experiments was an item emphasized under A2N-ISPI project. Table 4 also illustrates how the content was skewed towards technical issues. Monitoring and documentation of FFS activities was linked to performance in regard to handling technologies, as reflected in the curriculum. Emphasis was on how a variety yielded hence how resistant it was to pest attack, which is likely to have been used as an indicator for appropriateness of a variety. Reports focused more on technical issues, especially as they moved up the hierarchy from farmers at the lower level to donors at the highest level. For instance, the reports put accent on pest resistance and yield of the different varieties. How farmers interacted with the different technology components was often ignored, yet technology is developed in effort to improve upon agricultural production and rural or farmer livelihood. This implies that any technology that does not lead to any change in the way people live or make decisions is not worth being introduced to a community. More of the documentation of FFS is discussed in chapter five. Table 4 below offers an example of the contents of training for facilitators under the CIP-IPPHM project. This was a 3-day training prepared and conducted by the sweet-potato programme for all extension staff facilitators

Table 4: Curriculum for training of facilitators under CIP-IPPHM.

Day	Content or topic covered	Time allocated for a given topic	Source of resource person
Day One 08:30 - 17:00	Sweet potato variety development	08:30 - 9:30	NAARI - NARO
	Sweet potato agronomy	9:30 - 10:30	NAARI- NARO
	Break	10:30 - 11:00	
	Mineral nutrition	11:00 - 12:00	NAARI- NARO
	Tissue culture	12:00 - 13:00	NAARI- NARO
	Lunch	13:00 - 14:00	
	Tissue culture (tour of tissue culture lab - practical)	14:00 - 15:00	NAARI - NARO
	Crossing block (practical)	15:00 - 16:00	NAARI - NARO
	Sweet potato agronomy (practical)	16:00 - 17:00	NAARI - NARO
Day two 08:30-18:00	Seed multiplication	08:30 - 09:30	NAARI - NARO
	OFSP for vitamin A	09:30 - 10:00	NAARI - NARO
	Break	10:00 - 10:30	
	Sweet potato pest management	10:30 - 11:30	NAARI - NARO
	Sweet potato disease management	11:30 - 12:30	NAARI - NARO
	Sweet potato disease management (practical)	12:30 - 13:30	NAARI - NARO
	Lunch	13:30 - 14:30	
	Sweet potato pest management	14:30 - 15:30	NAARI - NARO
	Visit to farmer's field (neighbouring village)	15:30 - 18:00	NAARI - NARO
Day three	Sweet potato post harvest management	Morning	KARI - NARO
	Sweet potato post harvest practical session	Afternoon	KARI - NARO

(referred to as core trainers) in June 2002 at NAARI (Uganda). The table indicates clearly the one-way teacher-student interaction. Note the lack of discussion sessions and of sessions on skills in analysis and management of social interaction.

Mancini (2006) similarly reports on the skew towards training on insect ecology and bio-control principles in Indian rice FFS, with strong implications for the way in which FFS programmes subsequently function. Note that in both the sweet potato case cited above and in Mancini's example the curriculum was developed by scientists whose interest lay more in technology than in social issues, under an assumption that technologies (by themselves) provide solutions to problems in farming communities. Both cases appear to have lost sight of the idea that farmer knowledge enables better choices in regard to appropriate technology packages or components and delivery in local contexts. The challenge lies in aligning what farmers actually do with a given crop and what projects have to offer in relation to the same crop. Farmers' field realities or contributions can then be integrated with the researchers' views.

Inadequacy in training content exposed facilitators to the risk of being less relevant in some cases due to their inability to handle realities in the farmers' fields. Disease management was sometimes left out of training (though not in the example cited above), yet was a frequent occurrence in farmers' fields. Focus on agronomy and insect pests made facilitators somewhat inflexible, and at times they had little or nothing to say on disease-related problems at field level. Where (as above) efforts were made to include disease management the time allocated was inadequate for facilitators fully to understand pathology. Disease is a complex issue and often needed more time and practice. During my participation in some of the FFS sessions (AESA) in ISPI FFS in Tororo and Busia I realized that sometimes facilitators misguided farmers due to lack of information, or just ignorance on specific diseases of groundnuts, cabbage and maize (as will be more fully narrated below). Ability to scan and analyse what goes on in farms is a key skill for effective facilitation in technology extension.

In one Umoja FFS study plot neither the facilitator nor the farmers could recognize rust/burn disease that affected cabbage simply because it was very common for cabbage to get such lesions during a dry spell. All farmers could say was that "...these things come on the crop when it is very dry; we are used and it is normal..." was the response. In this case, provision of disease tolerant and disease free planting materials is important. Emphasis on and commitment to inculcate skills in regard observational learning and practice, thus enhancing analytical and diagnostic skills - a basic idea behind FFS - ought to help farmers to recognise the main local abnormalities due to disease and not to mistake disease for something else (a normal condition associated with dry weather).

In a second situation, as observed in the present study, a facilitator misguided farmers about the disease resistance of one improved groundnut variety Serenut 2. "...we were told that because the variety is resistant to drought and pests, there was no need to spray..." one of the farmers at the site told me. The facilitator had assumed these farmers were familiar with and interested in groundnuts. Groundnut was not, in fact, a very important crop in Busia and Tororo, and this probably explains why farmers did not actually pay much attention to what was happening, and were happy to let the crop "fend for itself". The facilitator did not 'read' the local context very carefully, and thus was unprepared for farmers taking his words to imply more than had been intended. The crop was lost to disease and effect of a leaf miner insect pest. The varieties were resistant to drought and effects of the aphid.

The miner, when Serenut 2 was first bred and released as a resistant variety, was probably a minor pest. The facilitator was not informed about the leaf miner and surprisingly the farmers did not know it was an insect affecting the crop! I moved with two of them to the plot and checked closely with them. We discovered some insects rolled behind the leaves. This suggested an absence of analytical observation concerning why farmers had been reluctant to cultivate groundnuts, resulting in a failure to devise appropriate ways of promoting interactions of farmers with the new or improved technology of improved groundnut varieties within the FFS framework.

In Busia, the study crop Maize (Longe 5) under ISPI was attacked by Northern Leaf Blight Disease (NLB). The improved maize variety Longe 5 was susceptible to NLB especially when planted late in the season. The facilitator and farmers did not know what was happening and could not do anything! Funnily enough, the neighbouring field of local maize was performing better, having healthy, darker leaves and taller plants. In one FFS under SPUH, where the study crop was cabbage, farmers under the instruction of their facilitator applied mulch to a plot that appeared to be doing better and left other plots, yet the treatment was spacing. The mentality of the extension worker (facilitator) appears to have been "better to do something rather than admit ignorance or failure". But inability to recognise what we do not know is what FFS was set up to overcome (as a discovery based learning model). Misplaced activism not only biases outcomes but is likely to lead to a wrong interpretation of context. Realities in the farming system on the ground are hidden by what facilitators presume to be the case, and this presumption serves to undermine discovery, instead maintaining the basic logic of the conventional transfer model (these people suffer because they do not yet have what we have or know).

Participatory approaches and techniques, hence facilitation and communication skills, needed in dealing with a group are a weak link in training of facilitators for FFS. In their study of extension workers in Ethiopia, Belay and Abebaw (2004) revealed communication and theory orientation as two of the weak links limiting the effectiveness of extension workers. Inadequate communication skills aggravate an inability to analyze their audience. The same situation applies to extension workers serving as facilitators in FFS projects in Uganda. This is why Rollins *et al.* (1994) agreed that personnel charged with informational and educational responsibilities also have information and education needs. Whereas technical competence is taken as very important, competence in analyzing social aspects of the farming system to be improved through FFS seem equally critical, yet largely missing. The technical content can only be translated into a farmer-related context through a language that suits farmer realities. Non-judgmental attitudes and ability to interact with different people are other qualities required in facilitation. To be able to understand others it is imperative that the facilitator understand himself and how his behaviour is likely to influence what farmers do or say (Kibwika 2006). A discussion with some project implementers led to recognition of the need to include facilitation and communication skills as a topic during facilitators' training. The challenge, however, is identifying people with adequate competence both in facilitation as a process of communication and also in the social analysis of group dynamics. Balancing technical with social analytical and communication skills is important in training or re-orienting agricultural extension.

Time allocation to sessions/topics
For effective facilitator training, time is an important factor. I happened to participate in two farmer facilitator training activities of two FFS projects as a resource person on facilitation and communication skills, not as a facilitator. Resource persons are not facilitators, and the reverse is equally true, although a resource person's role can be played in a facilitative approach. The time was very limited (at the utmost two hours per topic) yet topics to be covered were many, with many complexities in content. Facilitation and communication as extension-related or social skills, for instance, could not be handled effectively within the allocated two hours, especially as a single topic. In this situation, the lecture method proved a handy refuge.

In an effort to use the allotted time adequately I resorted to the lecture method and handouts, with very few group tasks. Little does this approach of training help the facilitators un-learn the conventional extension method of 'preaching' about recommended agricultural practices as opposed to encouraging farmer modifications that render technologies more appropriate. It also has a danger of suffocating the innovativeness or creativity of trainees. Based on my interactions with facilitators, I had prepared lots of things on interpersonal and communication skills that I felt were relevant for a facilitator but time was very limited. Sorting out what to give, how to give it, and in what quantity, was itself a challenge. The sessions were offered one after another. You could see the trainees tired but little could be done. The program had to be completed as scheduled. It reached a time when they just stared! In a tired state, how much concentration to engage with subject matter occurs? The brains are packed with too much information - a waste because it is never used. Translating such material into facilitation becomes a big dilemma. Provision of handouts does not necessarily guarantee that participants will pick up, read and make efforts to use the information therein. Acquisition of handouts seemed more a habit for many participants in training sessions or workshops. This was rather evident from the fact that some facilitators left the handouts behind upon ending the training.

Instruction language
The instruction language was English yet most of the farmers did not understand it well, though they could speak it. They needed more time to get to grips with the technical/academic language used by some trainers. "...I was very slow in catching up with the training because it was technical and in English which I could not cope with easily..." one of the farmer facilitators who had just returned from facilitator training course remarked. Processing everything learnt was very difficult for him. "There are words that I cannot translate from English to my local language, Ateso". This farmer facilitator did not get to grips with the manual. Again, it needed more time. Most trainers were not down to earth. Farmer facilitators' discomfort with English was revealed when they requested to make their presentations in their local languages (Swahili for Kenya and Ateso for Uganda) during one stakeholder review workshop in Busia (Blue York Hotel, Kenya). One of the participants translated from the local language to English for the benefit of the rest of the participants. This shows how the trainers also failed to analyse their audience, a factor in explaining why facilitators later tended to 'imitate' their trainers.

Training period
Training of facilitators was done prior to the start of the farming season so that by the time rains set in facilitators were well 'armed' and ready to work with FFS groups. Training during the off season denied facilitators chance to have hands-on practical sessions about the different technologies (crop improvement, soil improvement and pesticide use) in the field. But it was difficult to get extension workers and farmers off their stations during the rians because they were busy with various fieldwork-related activities. Some projects like MAK-IPM, under Makerere University, provided magnified pictures and/or photographs of different pests to help facilitators identify the different legume pests while in the field. From black and white

pictures it was difficult for facilitators to identify and distinguish between some insects in the fields. At some point facilitators were encouraged to take such insects to the entomologists in Makere for identification, but this did not work out given the time and money involved on the part of facilitator.

Duration of training for facilitators varied across projects and types of facilitators: longer for extension staff and shorter for farmers, as can be seen in Figure 4. Training duration of extension staff facilitators' decreased in later projects. Facilitators under the pioneer FAO-IPPM project underwent a thorough season-long, hands-on training lasting four months, but subsequent projects trained their facilitators for no more than two weeks at most. The four months training was intended to offer facilitators a thorough practical grounding in the technical aspects involved in cotton production with regard to agronomy and pest dynamics. Through practical hands-on sessions where every facilitator had his/her cotton plot, skills in cotton production were built, and this put facilitators in a better position to guide farmer-technology interactions.

The amount of time available would have been sufficient to allow facilitators also to carry out community analysis and study the social feasibility of the new technology. The focus and interest, however, was more with how the technology works than how society was likely to interact with the technology. The same weakness was encountered in other projects, but perhaps more understandably so, given the very much more limited total time available. In the cotton project training involved an action learning/research process approach where trainees (extension staff facilitators) returned home or work stations to try out what was taught about the (cotton) technology in question, then report back to the training centre to acquire more knowledge and skills, share challenges met with the rest of the group, devise ways of dealing with the challenges, followed by return to the work station to try out and implement ideas. The cycle continued till the end of the season. In this way, facilitators experienced a better internalization and understanding of what they were to train farmers about in the new

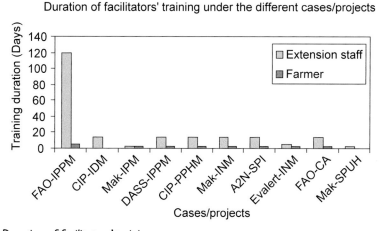

Figure 4: Duration of facilitators' training.

technology, but there was no specific space in which to "test" whether the technology fitted the local social and organizational contexts. This was the first time extension workers were exposed to practical training about a technology following the whole crop phenology from seed to seed. Despite lack of social content the training proved a valuable opportunity for most trainees, given that their previous knowledge was more theory-based than practical (skill) oriented. It is not untypical for Ugandan extension workers to have inadequate technical skills required in the field.

Subsequent limited training duration was mainly attributed to lack of funds. Additionally, it was also assumed that college training was adequate to the technologies being promoted under the projects, since these were not very new but only required recapitulation of a few aspects. For the case of farmer facilitators, it was argued that their involvement in the season-long FFS training sessions provided an adequate background concerning both methodology and technical aspects. However, the farmers themselves mentioned that one year of FFS sessions was not adequate to grasp the technical issues, especially related to pests. This was more of a concern in CIP-PPHM where farmer facilitators and farmers felt they could not find a reliable solution to the sweet potato weevil. The extension staff facilitators were expected to provide frequent back stopping to farmer facilitators but this hardly happened. They interacted less with practical cultivation aspects, and therefore were inadequately informed about most crop-farmer interactions, as compared to farmers who daily worked the soil. However, being busy with other things was often offered as an excuse for extension facilitators failing to back up farmer facilitators. This often appeared to be an excuse to avoid the potential embarrassment at exposing their ignorance about technologies about which farmers were better informed, especially in regard to local farming conditions and social context.

3.4 Mobilization of the farming community

In collaboration with local leaders at community level facilitators mobilized the farming community, as both individuals and groups, to work with FFS projects. Methods of mobilization were *ad hoc* [i.e. there seem to be no agreed method) and had implications for the type of people incorporated within the projects. Facilitators made use of their own networks, or exercised their own judgement about the types of persons required, and this was responsible for emergence of groups not fully representative of a cross-section of the larger farming community. This 'closed' mobilization tended towards less highly motivated individuals.

Mobilization started with involvement of administrators at district level. Researchers/scientists first met and briefed the district agricultural officers (DAO) or extension coordinators (DEC) about their intention to work with the farming community, with emphasis to a given methodology and/or technology. DEC then introduced the guests (researchers) to the district administrators heading the political (chairman of the district or LC V) and technical (Chief Administrative Officer - CAO) authorities. The researchers briefed the district administration about the project and emphasized the need for their support, which not only raised awareness but also interest in the project. Invitations to stakeholders at district level are supposed to be issued by the office of the LC V, but this was often delegated to the different heads of departments. This was the strategy that the MAK-SPUH project used when inviting stakeholders (those who sell, buy or advise on pesticides) to a two-day pesticide handling sensitization workshop. Mobilizing extension workers to work with the project was done by the DAO/DEC. This then had implications for which communities were to benefit from the project, given that the extension workers remained working in their areas of operation. Selected extension workers then took on the role of mobilizing farmers with the help of local leaders. The extension workers, together with the researchers, create awareness about projects and invite community participation in a meeting.

3.4.1 Mobilization by extension workers

FFS projects were mainly based on previous groups (Figure 5). Facilitators mobilized specific extant groups through chairpersons perceived to have a good, hardworking record in the village or sub-county. Such people serving as chair persons were generally termed contact farmers, progressive farmers, or model farmers who had frequently worked with previous projects operating in the area. Some held leadership positions in the village or were opinion leaders. Group leaders were more visible and had more resources to invest in such activities, compared to other group members (notably time and language skills). This gives the implication that mobilization is perhaps not the best route to a poverty-alleviation impact. Put another way, facilitators need more than people skills and technical expertise.

Buying-in a group leader into the project automatically meant buying-in the rest of the group members, since mobilization was done by the group leader. In Uganda, working with farmers in groups started in the early 1990's under the Agriculture Extension Project (AEP) with the aim of covering as many farmers as possible in the process of disseminating

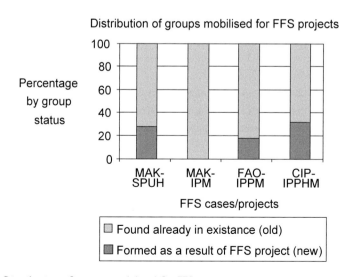

Figure 5: Distribution of groups mobilised for FFS.

recommended practices (MAAIF, 1995; CIET International, 1996). Groups have since been used by NGOs and others to to provide relief food, protection, and extension services (for both agriculture and health). The existence of groups in a district has had implications for NGO work. Since the establishment of PMA in 2001, it has been government policy for farmers to form groups in order to access services more easily. Facilitators selected prospective FFS groups prior to onset of rains. Group selection was therefore subject to facilitators' awareness of existing groups, interests, established relationships with group leadership and convenience (in distance, accessibility, working relationships and other aspects). The word "group" has become almost synonymous with "group leadership". It was more convenient to work with already existing groups (as illustrated in the bar graph), whether these earlier groups were self initiated or formed in response to (previous) project initiatives (Table 5), although new groups were formed in some areas without earlier groups. Whether working with pre-formed groups is an advantage, or smuggles in old habits contrary to the spirit of discovery-based learning remains to be examined.

Existence of farmers in a group implied recognition, interest, will, and need for collective effort in development activities. Place *et al.* (2002) support the view that groups tend to build their experiences best by taking on new activities rather forming and re-forming as new challenges arise. Following theories of stages of group development as outlined by, for example, Tuckman (1965) and Tuckman and Jensen (1977) older groups are expected to be more stable. Stability implies (according to the model) better performance, and better relationships among members. But in the present case it seems to matter more whether the group was self-formed or induced by previous development activity. Self initiated groups, especially under articulate leadership and sharing a common vision, seemed to have attained stability faster than some older "induced" groups, and hence offered better prospects for in-coming projects like FFS. Shared vision increases interaction and support for common objectives. High performing groups are said to have more consensus building and fewer conflicts (see Jehn, 1995; 1997; Jehn and Mannix, 2001).

Table 5: Formation of existing groups.

District	Source of group formation initiative	
	Self initiated (%)	Project initiated (%)
Busia	73	27
Tororo	41	59
Soroti	12	88
Mukono	31	69
Kiboga	47	53
Average %	40.8 =41	59.2 = 59

A factor not taken into account, however, is whether group formation selects against the poorer and less able sections of rural society. Projects like FFS come in a rush and typically have a short life span of 2-3 years to show results to bosses and donors. Because of pressure to show the worth of the project there is a growing tendency to prefer working with progressive farmers (groups and individuals). Such farmers are more knowledgeable, aware, resource endowed, responsible, and accountable, and quicker to try out new technologies. This is in agreement with findings by MFPED (2003) showing that agricultural extension workers are biased towards the better-off farmers, and forced to select above-average farmers to work with because of government emphasis on results, a situation that consequently limits access to services to a few farmers, as pointed out by MOPS (2001). Any one would fear to join a low-performing group for fear of being evaluated as a non performer too. The limitation of the 'hardworking group' syndrome, however, is that most subsequent incoming projects tend to work with the same narrow range of "elite" groups, leaving the majority with the faint hope that this small elite will somehow be public-spirited enough to take time off to disseminate useful information to the rest. Channekling activity towards the same narrow set of successful groups risks creating two extremes of very knowledgeable and much uninformed categories of farmers in the community. This may consequently create a sense of a class divide in which the "not knowledgeable" fear to approach the "knowledgeable", seen as people set apart (and perhaps superior in their own eyes).

In some cases, some (extension staff) facilitators selected malleable groups, i.e. groups willing to accept whatever they suggested. This undermined farmers' interest. Such groups were easier to work with through instruction, were uncomplaining and fearful to challenge the facilitator, therefore easily abused or cheated. An example was when there was a disagreement between one farmer facilitator and one extension staff facilitator (in the same sub-county) about selection of farmer groups by the farmer facilitator in a village under CIP-PPHM. Each facilitator was supposed to identify his own group to work with. But in this case the extension staff facilitator selected a group for the farmer facilitator and told him not to select a different group. The farmer facilitator was not comfortable with this, because according to him, members of the group were not hard working, and the groups was too far from where the farmer facilitator lived. But to the extension staff, this was the only option, perhaps because of personal interests. The program assistant and I learnt about this when we went to verify the presence and readiness of groups, as we prepared to deliver planting materials to the various groups under the project (CIP-PPHM). We thought the farmer facilitator should go ahead and pick a group of his own. Later, the farmer facilitator mobilized formation of a group of his choice and convenience. He managed to persuade two old IPPM-FFS groups to come together and form a new, single IPPHM-FFS group. In this case, the farmer had some information about the characteristics of the people he was to work with in terms of their interest in the work, though probably not whether they were interested in the technology to be introduced.

3.3.2 Mobilization of farmers by local leaders

In areas where there were no groups to the knowledge of the facilitator, local village leaders took the lead in ensuring formation of farmers' groups. Although local village leaders helped mobilize communities to form groups, the degree of objectivity varied. Some clearly explained the objective of FFS projects to improve farming practices. Others, however, emphasized availability of and access to funds as a lure. To inform people about the SPUH-MAK project, posters were pinned up around trading centres, announcements made in e.g. churches, at funerals, and at times on the radio to invite interested people. Some local leaders walked from home to home inviting village members to a meeting to discuss the project and the need to form a group. The local leaders walking from home to home, however, tended to inform only their close friends and relatives, leaving out other people. Again it is clear that the methods used in mobilization had implications for the type of people who then joined the new farmer groups. Open invitation offered more chance of enrolling genuinely interested members while closed mobilization risked an influx of less interested members who joined because they were told to do so, and therefore expect a pay off, or were under pressure of obligation to the person inviting them.

If results of local leadership involvement in mobilization are so far mixed, recent proposals (presented during the 2005/6 Uganda budget exercise) for government to provide salaries to such local leaders look set to change the way the entire process works. The interests of government may now take precedence over community interests, especially given that Uganda is currently nurturing a multiparty system in which almost all local leaders at the village level are likely to pledge loyalty to the current ruling party (the National Resistance Movement, NRM).

Through their community mobilization role, local (political) leaders used FFS projects as an opportunity to reward specific colleagues[17]. In many cases, leaders at village level became chairpersons of the new groups, a common phenomenon where one who brings up an idea presumes to take the lead. Issue raisers may not necessarily have the patience, commitment or will to take lead in running development activities. As one way of ensuring commitment of group members, the interim[18] group chairpersons levied membership fees as a basis to open up a group account, as required by FFS projects. Some farmers were unable to pay the fee and could not continue to be part of the group, therefore. One argument in favour of a levy is that people tend to attach more value to matters calling for their financial commitment. However, in some cases, the results may be counter-productive, especially where the expenditure is no

---

[17] This category included people who helped in mobilizing votes for them (commonly called campaign agents), relatives, and close friends. Promising involvement of others from the electorate in similar subsequent projects was a strategy in some cases of ensuring another term of office in a leadership position.

[18] In situations where there were no groups, a community or village person who took up an active role of encouraging colleagues or community members to form a group assumed the role of the leader (or chairperson) of that group. Experience from this study revealed group members of such groups often elected and confirmed the group 'initiators' as chairpersons of such groups. These chairpersons were the links between the group and the extension worker or project personnel.

real hardship. Often, those in a position to pay are not the 'right' farmers in terms of agro-technology needs.

Doubts due to former bad experiences with false projects demoralize farmers, and reduce their active commitment to a new project. Such a situation is a challenge for local leaders in the mobilization process. In one of the villages it was difficult for facilitator and local leader to convince farmers about the MAK-SPUH project because they themselves had no proof that the project would come to reality. "As soon as I left Makerere, where I had attended the training on pesticides, I started 'selling' what I did not know..." said the facilitator. Farmers in this village (Kakunyu-Kiboga) had experienced people approaching them in the name of development projects, soliciting money from people to become members, and later failing to turn up. Farmers quickly lost trust in the leaders who mobilized them for this 'unrealized' project. The facilitator could not clear farmers' doubts, but did not want to take on the blame for failing to form a group. Based on farmers' previous experience, there was fear of losing more money in similar ways. Failure of the project team to turn up after fixing an appointment with the communities to inaugurate the program only worsened the situation. Depositing money on the farmers' account, purchase and provision of experimental materials to groups and finally the effort of the team to visit the groups provided some hope and motivation among people to engage in project activities. Provision of money and materials to farmer groups alone is never satisfactory. Interactions with them via visits and informal talks make them feel their worth as social beings too. It is better to wait, than to make farmers wait, because then the chances of killing their interest and commitment at the first time of asking are minimized. As the saying goes 'there is no second chance to create a first impression'.

3.4. Effect of mobilization method on group size and dynamics

Studies conducted of the Ugandan NAADS initiative and farmer groups (Kayanja, 2003; Obaa, 2004) indicated that presence of a project in a community tends to create an upsurge in numbers of groups and in group membership, which later reduces when members fail to realize their interests and expectations. Although NAADS is a national program, and FFS a government-approved international agency/NGO initiative, the trend is similar. Under FFS, use of groups is a prerequisite for a community to be part of an FFS project. Following experience from Asia, a typical FFS group is composed of about 25-30 farmers (van de Fliert, 1993; Kenmore, 1991). During the mobilization exercise, bigger group sizes were encouraged, with the expectation that numbers would stabilize around 25 as some members fell off along the way. Thus new groups started with a typical size of 25-37 members. Most (more than 80%) of old groups, by contrast, had less than 20 members, with a majority in the range 10-15. As a prerequisite to register with an FFS project, old groups had to increase group membership to 25, and ensure gender balance, which was done with some hesitation, where the old groups were purely female in membership.

Encouraging larger groups, as often promoted by donors and NGOs, is likely to create trade offs between economies of scale and group cohesion (larger groups are less cohesive), yet cohesion is a critical factor in interaction of group members (Stringfellow *et al.*, 1997). Although enlargement was one way of ensuring inclusion of more people (females as well as males) in FFS projects, it interrupted group dynamics: freedom of women was minimized in the presence of men. Women-only groups expressed their discomfort and fear of being bulldozed by men. "Men are bad people to work with because they want to own everything and will want to be leaders in our own group..." a woman from the Asianut FFS group stated. To some groups, non-members were seen as lazy and not interested in collective work. So they had reluctance to expand, fearing an influx of less committed members. Various ruses were used to get the required list of 25 group members while still basically remaining with the original group of 10-15 members. Although the lists showed totals of 25 members, most new members were not active and did not turn up for group work. This was because they did not choose to join, but the existing group members just decided upon names to add to the list to make it up to 25. To some groups it was a responsibility of the executive, while in other groups all members had the responsibility of bringing in new "sleeping" members.

New members on the list were either old people, relatives of established group members within the same village or husbands of women in established women's groups. Two groups registered school students, who were never actually available. New "sleeping" members did not fully understand the objective of the group and were not very free in the group. As a result of difference in focus, one group had 2 sub-groups with different objectives. The pioneer group had frequent meetings and members cooperated closely, while the "big" FFS group met once a week. There was a group in Busia that later divided in two, with pioneer members accusing new members of laziness and less involvement in group work. Such situations kept de facto membership down to 10-15, a number actually found convenient by some facilitators (especially farmer). "It is easier to work with few people in a group..." The smaller the group is, the greater the interaction and active participation of individuals. In bigger groups it is easier for the shy or less committed to hide in the background. It is worth noting that these points about *de facto* group size are not limited to Ugandan FFS. In Java (Indonesia), where FFS started, Winarto (2004: 146) found out that on average only about 14 farmers per group were active per group. Some of the reasons why farmers join and leave projects are given below (Box 6).

A range of other reasons were given by informants for reduction in active group membership. One set of factors included a preference for other crops for commercial purposes than the targeted/ study crop. Most farmers in FFS groups under MAK-SPUH preferred cassava, maize, groundnuts, onions and tobacco to vegetables. For example only 5 farmers in one of the groups were interested in and serious growers of tomatoes and cabbage (the target crops of the project); under FAO-IPPM there was more preference to groundnuts than the targeted cotton, and for millet and groundnuts for groups under the sweet potato-focused CIP-IPPHM. Precedence of individual/home fields over group fields, because participants were sure of enjoying the benefits individually also figured among reasons for reduced commitment. "Spending the whole afternoon in my garden pays more than wasting it in FFS training...", said one informant. Others complained about unrealized high expectations of acquiring handouts and a failure to understand trainings, due to the academic nature of the

---

Box 6: What lured farmers to join and persuaded them to leave FFS groups/projects?

Framers attributed eagerness to become members of active FFS groups mainly to various perceived benefits (realistic and unrealistic) and when these proved hard to realise disappointment or frustration contributed to decline in group size and divided interests of members. Reasons why farmers wanted to join FFS groups/projects included desire to have bigger and better looking gardens (mainly modelled on the collective commercial fields), realizing better yields to provide food and generate income for their families, and opportunity to eat something when in FFS. This last was mentioned by members in Kawabona-Kabosi FFS that later joined A2N-ISPI. FAO-IPPM provided money for snacks at the beginning (1999-2000) but these payments were later scrapped. No other applications provided any form of snacks. To some farmers the desire to stay longer in the field (from 6:00am to 2:00pm) implied more learning. The belief was that the longer the time spent collectively together in the field, the more people learnt how to improve on yields. "Those people learn a lot and do much because they stay in the field together for a longer time" said one farmer.

Others had the expectation of getting free pesticides and hoes. One man quit FFS after realizing that there were no inputs given out. "…if they are all digging using their hands and hoes like me without any help from the project, why should I waste my time in the group…?" Others had the hope that getting FFS training was a 'ticket' to get loans. "…I hear from those who were once in field schools that at the end of the training, FAO was supposed to provide loans to the trainees…" Such members lost interest and abandoned the group on realizing that the rumour about FAO providing loans was not true. To the majority of individuals and groups the aim of accessing money from the project was the target - and the group development fund a sign of hope. The examples above indicate that most farmer were mainly interested in short term rather than long term benefits, and looked more to 'to be given' than 'to work for'. Perhaps they did not yet know that that knowledge can make a difference or perhaps they just gave up. It remains an issue to know how to instil a spirit of being proactive.

---

process (measuring heights and lengths, counting, questions asked during presentations) and, finally, the personality and training methods of unpopular facilitators. The more friendly and interesting the facilitator, the more members turned up. The nature of topics handled also had implications for regularity in attendance and therefore involvement in FFS activities. When all the above factors are taken into account it is easy to see why in most FFS groups management of experimental plots was mainly carried out by the chairperson of the group and a small group of no more than 3-4 keen farmers (mainly executive members). The chair persons hosted the experiments and cared for them for the good of the group.

Why were groups of 25 seen to be so important? The issue of group size seems to have been taken for granted because the explanation given was limited to the requirements of the project. Facilitators did not understand any functional logic behind 25. It was simply a project norm - a rule to be applied. Even in situations where membership was more or less (which was the situation in most cases) they frequently reported a membership of 25! Yet, despite the list, active members accounted for only 10-20% of the group and tended to have a history of involvement in older existing groups. More (60-100%) participated in collective work of opening up land, planting and harvesting, but then the numbers gradually reduced with time. Facilitators, however, were more comfortable with smaller sized groups. According to one facilitator "the bigger the group (more than 15) the more uncooperative the farmers..." Members fell off especially after realizing that their expectations were far from being attained and when they developed conflicts in interest. Exaggeration of group size was partly the result of the mobilization techniques used by local leaders and group leaders. The feeling that reporting a higher number of members per group justified more funding from the funding agencies was mentioned by some facilitators in Kenya. "Mentioning a smaller group membership has implications for funding. The funders want to work with big groups...", said one district coordinator, defending the group membership norm of 25.

Facilitators' interest in quantities seemed to align with their attitude towards adoption (how many farmers are involved?) as opposed to provoking analysis of reasons for (non-)involvement. Given the different prevailing local circumstances, adaptation (therefore modification) to suit farmers would seem a better stance than a focus on adopting (where farmers are expected to fit within the rules or context of a "pre-cooked" technology). Of course project administrations use figures on numbers of farmers reached to justify their interventions and signal success. But in addition to the quantities trained, it would have been helpful if projects had followed up the qualitative part of what happened as a result of the training offered. How farmer-technology interactions continue, and why? This would give a better opportunity to identify areas where projects lack adequate information concerning technology-society linkage. A learning organization (as FFS claims to be) would use careful observation and interpretation of what farmers do with technologies as a rich source for improving technology development and implementation and operation of extension systems more generally. This study found little evidence that these kinds of learning modalities were in place. Facilitators had simply not been encouraged to understand that a well-analysed "negative" report would be more helpful to project administrations than bland reports of alleged success.

3.4.4 Verification and inauguration of projects in the communities

Inauguration activities indirectly served as strategies to mobilize and 'sell' the project technologies to more people among the farming community. Verification of groups helped identify irregularities that might jeopardize the operation of the project and therefore derail interventions. This was mainly related to presence of a group and readiness to take part in the project activities. In the process of visiting participating villages to verify existence and willingness of groups to work on FFS objectives, project assistants (FAO-IPPM and CIP-PPHM) explained the objectives and scope of the project to damp down over-excited farmer expectations. However, verification was not a regular exercise since a high degree of trust was invested in facilitators. During the verification exercise absence of groups in some cases, and un-willingness of some members to work with the project, were discovered. Verification of groups served to weed out some "bad apples" and minimized waste of project resources.

Under FAO-IPPM, one facilitator failed to mobilize farmers. He therefore did not have a farmers' group to work with as a facilitator but was expected to submit membership lists and the name of his group to the FFS secretariat at the district level. Provision of money, experimental materials (improved varieties/seed, fertilizers and pesticides) and stationery (flip charts, exercise books, pens, crayons or markers) for FFS sessions took place only once a group had been identified and details submitted to the secretariat. This facilitator developed and submitted a list of names. These all happened to be in-mates in a jail in one of the remoter sub-counties. His interest was to get access to FFS materials for his own purposes. There was no way prisoners would be released from jail to attend FFS activities.

Verification brought to light some further evidence of similar situations, indicating some degree of false reporting of projects in terms of area, number of groups and individual farmers. This checking is commonplace with any kind of rural development work and some cases of fraud are almost inevitable. But the notion of verification was rather limited in scope. It did not extend, for example, to verifying whether the technology introduced was feasible in the local context. Furthermore, it rather undermined some basic assumptions of FFS. If the right community of learners is targeted, it ought not to be necessary to carry out a verification exercise to discover whether groups exist and are ready to receive new technology. The groups would verify themselves through their eagerness and activism in FFS activity. Attempts to check and weed out fake groups also nullified the assumption that extension workers' had a valuable understanding and knowledge about farmers and local contexts. A rogue facilitator and fake group ought not long to withstand the scrutiny of peers.

For CIP-PPHM, the objective was not to verify but to check whether the farmer groups were prepared to receive planting vines about to be ferried in a week's time. It was expected that groups had already prepared fields (about 0.5 acre) for the experiments. "I do not want to look a fool when these people come and find no ready fields..." one facilitator mentioned. Vines were supplied by a research institute (NAARI) in Kampala. It transpired that not all groups under extension staff supervision were ready to receive the vines, even though facilitators were expected to have informed (and some did inform) their respective farmer groups. Farmers' lack of preparedness, despite being warned, spoke volumes about their lack of enthusiasm for the technology package (especially improved varieties, and planting methods), an important indicator of the inappropriateness of the technology, or the lack of importance attached to sweet potatoes in these villages. A factor in some cases was that extension facilitators were busier with NAADS activities, which paid better and were less demanding (instruction was oriented around recommended agricultural technologies and practices). Engaging with farmer-technology interactions in FFS required patience and time that extension workers did not seem to have or felt uncomfortable supplying.

One group almost disintegrated because of the absence of their chairperson, who was chronically absent from home, attending bible study courses in Kenya. The facilitator never verified this problem, or even talked with any of the members, but simply assumed that the pastor had 'converted' all his members to FFS activities. Clearly, the facilitator was assuming that without its chairperson this group ceased to be, and that once a specific individual had been appointed to lead no deputy could be appointed. In fact, other members were reluctant to host the project because of the responsibilities left to the leader - to provide land for the class and for the experiments. Where chairpersons did offer part of their land/fields for group activities, selfish reasons played a part (the attraction of claiming outcomes of experiments). The group was supposed to agree on how to share the benefits, but how to reward the person who offered land was often overlooked. Under MAK-IPM, such cases were catered for by providing a fee to hire land. But elsewhere unresolved tensions over this issue undermined FFS experiments. In western Kenya, one farmer hosting project activities destroyed group experiments, after she began to feel that she did not benefit from her land. A similar situation was also observed in Soroti.

Inauguration was done by MAK-SPUH project, but in the course of the FFS activities. Efforts to meet the groups earlier were fruitless due to the heavy work schedule of the Makerere team leaders, given that they were full time lecturers. During the inauguration, the MAK-SPUH project team interacted with the farming community (FFS groups) and local leaders in the area. Issues regarding safe pesticide use in vegetable growing, fund allocation (for transparency), and collective learning were addressed. This was an opportunity for the Makerere team to interact with administrators at sub-county level (extension worker, the LCIII, sub-county chief and councillors) and allay fears about 'false projects'. During the inaugurations, farming communities raised a number of issues (Box 7) that project implementers were not to ignore.

Points raised implied that farmers faced multiple problems going beyond what projects offer single-handed, where solutions need collective effort. The project team put more emphasis on the training it had come to offer. Avocado, reduced pesticide prices and market assurance were deemed "beyond the project". But how can a "participatory" project strike a

---

Box 7: Some issues raised by farmers during one inauguration session in Kiboga district.

- The need for frequent and constant communication with the project team
- Desire to have a female facilitator work with women once in a while
- Interest in learning when and how to spray, given that most farmers in the area produced tomatoes and cabbages, which where the heavily sprayed crops with a commercial value
- Desire to learn how to produce and handle other crops apart from the study crops of the project (tomatoes and cabbages). This was a request from one farmer who was interested in avocado because his small plantation had problems
- "Can we get pesticides at reduced prices?"
- How to ensure market of the produce and avoid flooding the market with the crop
- Assurance from the project team that it was serious. "We get people from outside telling us to form groups but after we have formed the groups, we do not hear/see them. At the end of it all we fail to get to know what is going on. How different will you be from such people/projects?"

---

balance between what it has to offer and what farmers express interest in learning more about? Leaving farmers in suspense because current projects cannot answer some of their problems is one way of 'chasing' them away, and might make them resent future projects, feeling they dimply do not listen. However, it is hard for projects to take on issues raised in communities owing to fixed budgets and time frames, specialized expertise and pre-ordained focus on specific technology. One option to consider is whether any current project might serve also as a "broker" by seeking out and linking up with other partners in position to work on issues beyond the current mandate. Some could even be handled as special topics in the FFS group. However, to avoid wasting resources on an enterprise in which very few farmers are interested, FFS teams also need to develop a simple methodology to assess the extent to which these community expressions of interest reflect genuine communal concerns.

3.5 Concluding remarks

FFS in Uganda was oriented more towards technology in stock rather than discovery-based learning targeted on in situ innovation or adaptation. This fitted the top-down instructional biases of the actors (agricultural professionals) involved in research and extension. Besides lacking time and resources for in situ technology development, the actors had not been trained in participation, and as a result end up offering what they were used to supplying in the name of participation. For participatory approaches to be conducted adequately and effectively there is need for the actors to be trained in techniques of participation. This also calls for institutional creativity with regard to new or innovative ways of actively engaging and supporting farmers for technology selection, development and use. Since many technologies "imposed" on farmers do not work for them, an alternative would be to encourage local innovation processes. Such processes exist (e.g. numerous storage and food processing techniques) and need to be fully assessed for the scope they offer FFS initiatives to "add value". Identifying scope for such innovation activity, however, calls for an understanding of the local system and good analytical skills on the part of FFS functionaries.

This chapter has also shown that a "top-down" bias has also gripped FFS in Uganda with regard to selection and training of facilitators. Bias towards technical issues (recommended practices) created a situation that reinforced instructional methodologies in which farmers are told the "right" things to do. This denied facilitators the opportunity to be equipped with adequate skills to understand and analyse the local system and to introduce interventions suited to local situation/contexts. The mobilisation approach so far used in Uganda needs to be rethought, if the right people and more suitable interventions are to be introduced to specific communities. Focus on elites as the most common way of mobilizing farmers raises a fundamental issue about the inclusiveness and relevance of introduced technologies to the farming system of the targeted community. Although it may result in more effective initial use of limited time and resources, given that elites are often better informed and reliable in attaining results deemed acceptable by projects, it tends to result in the marginalization of a large group of people, the non-elites. Outcomes from elite-oriented projects may prove non-sustainable, since elites enjoy resources not available to the majority, or keep (new) knowledge and skills to themselves. Effective mobilization requires an understanding of agro-ecological dynamics, and also social relations and politics among farming communities. Knowledge of interaction patterns would help mobilize the right people in appropriate ways. It is not hard to conclude that FFS in Uganda lacked social analysis.

Thus we have seen that in Ugandan FFS projects, technologies were pre-manufactured and introduced top-down. Communities were involved neither in choosing technology packages or components. Investigations for feasibility of new technologies were hardly undertaken prior to their introduction via FFS. We now need to examine the issue of the extent to which these technologies suited the contexts in which they were applied. The next chapter approaches this issue from the perspective of the communities of learners.

CHAPTER FOUR

FFS performance in wider context: farming and social systems in eastern and central Uganda

4.1 Introduction

In this chapter the technographic approach to Farmer Field Schools (FFS) in Uganda brings us to an analysis of the response of farming communities to FFS interventions. The farming communities analysed here mainly included five FFS groups[19]: Abuket FFS (Abuket village, Kyeere sub-county, Soroti district); Sihubira FFS (Sihubira village, Lunyo sub-county, Busia district); Karwok FFS (Karwok village, Molo sub-county, Tororo district); Mwinho Akuwa Tweyambe group (Buwolomera village, Bulamagi sub-county, Iganga district); and Kiddawalime farmers group (Mulagi village/sub-county, Kiboga district). Information in this chapter, however, was not limited to the five FFS groups but also included some observations from other FFS groups visited (Annex 3) during this study. From our understanding of learning as social practice, as explained in chapter one, groups of farmers are understood here as communities of learners. Technological change, defined in chapter three as additions to the stock of instrumentalities and knowledge, always incorporates learning. Farmer Field Schools create a situation where introduction of external technological inputs is combined with cross-contextual learning, following Lave's (1995) model of teaching. What comes out of this chapter is that farmers are indeed keen learners but that they do not, in all circumstances, accept technological inputs from external sources, and thus do not fully engage in the teaching process of an FFS. Moreover, the teaching (or training) method itself can lead to certain interactions among participants resulting in an FFS failing to reach its goals. To analyse FFS performance thus requires a technography of the teaching methods within FFS. The key question to be answered in this chapter, centred on the performance of different projects where an FFS tries to link to learning capacity with participatory set up, is how successful was the FFS in establishing such a connection?

The projects analysed are: Integrated Pest Management under Makerere University (MAK-IPM) implemented in Iganga and Kumi districts, focused on groundnuts and cowpeas; Safe Pesticide Use and Handling also under Makerere University (MAK-SPUH), implemented in Kiboga, Mukono and Mbarara districts, focused on vegetables, especially tomatoes and cabbages; Soil Productivity Improvement under Africa 2000 Network (A2N-ISPI) implemented in Busia and Tororo, focused on maize and groundnuts; and Integrated Pest and Post Harvest Management under CIP (CIP-IPPHM) in Soroti district, focused on sweet potatoes. All these crops are seen as important sources of food and income and therefore

---

[19] The arrangement used in distributing FFS projects was one FFS in a village and sub-county. Identification of participating communities began with the sub-county then came down to a village. With use of farmer facilitators, the number of FFS groups in a sub-county increased.

formed the basis for project interventions. In order to get a clear picture of project impact some assessment is needed of who the communities of learners are, what they do and how they relate to the interventions. Chapter two revealed that overall project objectives and mandates reflected the institutional hierarchy of national and international organisations involved in the promotion of FFS. But the extent to which these processes affected the situation on the ground remained unclear. This chapter, therefore, navigates through what goes on in the farming systems and the wider social settings occupied by communities of learners and seeks to document what the different FFS projects contributed to these systems and settings.

Farming engages people in a range of complementary activities for maintenance of livelihoods. A farming system encompasses all productive activities in which farmers (i.e. members of farming households) become involved. Every specific farming activity affects related activities and processes, resulting in an overall performance and a certain outcomes. For this reason, farming is often considered to have systemic features, and a proper account requires a systems description (i.e. farming systems research). Historically, farming systems research has aimed to understand what farmers do, how they do it and why they do it (Collinson, 2000). Farming systems analysis then forms the basis for defining and implementing appropriate and better farming options. Lynam (2002) notes, however, that the objective to understand the functioning of farming systems has at times become disconnected from insight into processes underlying farmer response to technologies at farm level. Understanding the complexity of farming livelihoods and system performance is particularly important in restoring this dimension of practical applicability (Cleaver, 2002). This chapter thus makes an attempt to encompass this complexity for the areas of Uganda mentioned.

4.2 Technology interventions and prevailing farming practices

In seeking to analyse performance of technologies, the present section adopts a case-by-case approach, with some emphasis on both system and social performance in order to bring out clearly what each project attempted to do and how communities and farmers responded. After a discussion of the positioning of FFS interventions (Section 4.3) social process is brought to the forefront in Section 4.4., and linked to technological performance. It should be noted that projects varied in number and types of interventions to which farmers responded. Therefore the analyses of various projects given in following sections differ somewhat in length and format. But a recurrent theme is the differentiation of gender roles. Farming activities are performed by men and women, but gender differentiation as encountered in the field was not reflected in the organization of FFS.

4.2.1 MAK-IPM project technology interventions - pest management

The critical issue addressed was to minimise cowpea and groundnut yield losses due to insect pests (especially the aphids), mainly through use of improved varieties and minimal pesticide use. Cowpea and groundnuts are among the most important legume food and cash crops in eastern Uganda, especially in Iganga and Kumi district where the MAK-IPM project operated. The two crops are heavily curtailed by a complex of pests and diseases that sometimes amount to 100% yield loss. The inferior pest resistance of local varieties is often seen as the major factor. Though preferred for their taste, local varieties are highly susceptible to aphids, the most important insect pest. Aphid-affected plants become vulnerable to rosette disease. As a result, farmers (especially the more commercially oriented) tend to use higher amounts of pesticides. Pesticides are expensive, however. So in order to minimise losses to pests and improve output of legumes, Makerere University developed an IPM package to be spread through FFS. The package consists of improved varieties and spray regimes (Table 6). The new technology aimed at minimizing production costs through reduced dependency on pesticide use. The objective was to create awareness among cowpea and groundnut farmers about the existence of cheaper alternatives. Planting at the onset of rains, early weeding, row planting and optimal or recommended spacing between plants were taken up in experiments run by FFS groups. These were constant factors. Spacing used included 40cm by 10cm for spreading groundnut varieties, 30cm by 10cm for erect groundnut varieties and 60cm by 20cm for improved cowpea. Whether the introduced spacing was wider or closer depended on the different farmers' practices but in most cases however, the spacing was wider than the traditional spacing.

For groundnuts, three improved varieties, Igola-I, Serenut I-R (spreading type) and Serenut II (erect type), were selected from a legume programme hosted in SAARI-NARO. SAARI, situated in Soroti district, is a research institute under NARO whose mandate is to conduct research (in animal and [20]crop improvement) suitable for the semi arid farming systems that include Teso. These three varieties showed increased resistance to aphids, tolerance of drought

---

[20] Crops under the mandate of SAARI include cotton, groundnuts, sesame, sunflower, pigeon peas, cowpeas, millet, sorghum and pasture. Other areas mandated to SAARI are improvement of local cattle, goat and chicken breeds and development of animal traction technology. The semi-arid agro-ecological zones for which SAARI undertakes research Teso, Lango, Acholi, Wesrt-Nile and Masindi.

Table 6: FFS experiments designed to address pest problems in cowpea and groundnut.

Treatments	Varieties	
	Improved varieties	**Local varieties**
No spray	Both legumes	Both legumes
Two sprays	Only groundnuts	Groundnuts
Three sprays	Only cowpea	Cowpea
Four sprays	Only groundnuts	Groundnuts
Farmers' practice	Only cowpea	Cowpea
Weekly spraying	Only cowpea	Cowpea

and early maturity. Serenut I-R was susceptible to rosette but tasted better. Farmers brought two common local varieties for comparison purposes. Improved varieties were big seeded when compared to the local varieties. Because of taste and better groundnut paste for making smooth sauce, farmers preferred their small seeded local varieties in spite of susceptibility to pests and drought. The improved varieties were less tasty and allegedly left some bitter taste in the mouth. The majority of groundnut farmers were subsistence oriented. Difficulty in accessing improved seed, due to limited availability and high costs, was a problem, and most farmers considered it not worth the effort to use the new technology.

For the cowpea experiments one improved variety, MU-93, and one local variety, *Ebelat*, were used. MU-93 was developed by a cowpea improvement project led by Makerere researchers. The main feature of this variety was pest resistance. In Uganda, cowpea leaves and grains are both used as food ingredients. Young tender cowpea leaves are plucked for relish as the crop develops. Removal of apical leaves reduces vegetative growth and stimulates flower development, resulting in more pods and grains. Plucking also results in a reduced canopy, with the effect that a micro environment is created unfavourable to some cowpea pests. However, the local *Ebelat* variety was very susceptible to pests, both in the field and in storage. But apparently this did not weigh against its better taste and processing features. The late maturation and relatively rough and unpalatable leaves of MU-93 discouraged farmers from taking it up. Its lack of soft edible leaves implied no plucking of leaves, development of a denser canopy that suppressed flowering and pod development. Low pod production and brown coloured grains were not appreciated. Local varieties are white seeded. Moreover, the new variety was a spreading type; necessitating row planting and use of stakes to support the plant grow upwards, a practice farmers were not familiar with, and demanding extra labour. Local semi-erect varieties are sown by broadcasting. Broadcasting saves time and effort.

Among the many experiments farmers did (Table 6) improved varieties and three sprayings (at budding/vegetative, flowering, and podding stages) addressed issues of high production costs due to expense of pesticides and low yields. Choice of these experiments was based on prior studies conducted under the cowpea improvement project to understand farmers practices in cowpea pest management (Isubikalu 1998) and on-farm experiments to identify promising cowpea pest control strategies (Karungi and Adipala, 2004) that were used to design an IPM package for cowpea farmers in Eastern Uganda.

IPM components, especially early planting (1-2 weeks after the rains), use of improved varieties, and three sprayings were largely taken up by commercial cowpea producers, probably because they had financial capital to invest in the cowpea business. Subsistence farmers rarely sprayed because of the expense involved, and their need to harvest leaves for consumption. While commercially oriented farmers were more interested in cowpea grain yield, subsistence farmers were more interested in the leaves. Reduced pesticide cost is only an incentive when farmers can afford such an input in the first place (i.e. it is of interest only to commercial farmers). The majority of subsistence farmers prefer local varieties for palatability and ease of seed acquisition, in spite of susceptibility to pests. The same trend was observed for groundnuts. Improved crop production practices were mainly taken up by the very small group of commercially oriented cultivators. To cater for the interest of the subsistence farmers, researchers at SAARI carried on and developed more dual purpose improved cowpea varieties Secow I and II for leaf (subsistence) and grain yield (commercial) (NARO, 2002; Emeetai-Areke *et al.*, 2004). In spite of the vegetative characteristic liked by subsistence farmers, Secow I and II, like MU-93, did not weigh against the local varieties in taste and processing features. It also became more suitable for commercial farmers. Because Secow leaves were less liked for consumption and rarely plucked, realising better yield necessitated slashing. Slashing is an extra cost that is likely to make farmers shy away from growing the varieties Secow I and II. Lack of funds has apparently affected continuation of research on cowpea improvement.

What this case shows, therefore, is that in taking up a new variety, farmers' interests focus more on edibility, including tenderness, appearance and taste, than on yield. Yield, seed size and resistance count only as secondary issues by subsistence farmers, yet researchers perceive them as the most important parameters. When introducing a new technology, it is important to know what counts most for the farmers meant to benefit from such interventions. The labour implications are especially important. In general, farmers tend to shy away from new technologies/varieties that demand use of more labour than applicable to local varieties.

2.2.2 MAK-SPUH project technology intervention - pesticide use and handling

In the case of MAK-SPUH, there was a focus on reducing pesticide use among crops where chemical control had hitherto been seen as the sole effective remedy for pests and diseases. In Kiboga, one of the districts in which MAK-SPUH operated, tomatoes and cabbages are a major source of income and frequently sprayed. Vegetable farmers frequently suffer from pesticide related illness. In California, Calvert *et al.* (2004) observed higher rates of pesticide-related illness among workers in the agriculture industry. As Dinham (2003) has observed, fruits and vegetables together account for the major share of the global pesticide market. In the present case the project sought to minimize exposure of Ugandan vegetable farmers to pesticides, and to introduce a variety of methods (Table 7) geared to safe pesticide handling.

In Uganda, farmers engaged in vegetable production are mainly commercial oriented. These farmers (mainly males) have more access to land in swampy areas where vegetables perform better. In most cases, the swamps are either part of their land or are hired purposely to grow vegetables. Land in a swamp is more valuable than land on upland because of its productivity throughout the year. The advantage of swamps is presence of moisture even in the dry season. Spraying was often carried out by casual labourers hired at a temporary basis. Farmers' practice of depending on chemicals (pesticides and fungicides) in vegetable production exposes them to health risks, a situation that MAK-SPUH attempted to address through FFS.

Farmers' practices in relation to safe pesticide use project interventions
Some farmer practices favoured incidence of blight and hence increased production costs due to higher pesticide use in attempts to control disease. Farmers used local (small fruited) tomato varieties in the belief that 'tinned' tomatoes had low viability. Local varieties were smaller and more susceptible to diseases such as blight. Previous harvests provided seed shared

between neighbours, as fruits, seed or seedlings. Although seed acquisition was cheap, chances of spreading seed-borne blight were high. The local tomatoes were mainly planted directly in the field at an estimated spacing of 90x90cm. This was wide enough to allow creeping of the stems, and thus adequate space for fruit development. Farmers often planted tomatoes in the same place year after year, a situation that encourages build-up of disease pathogens in the soil. Spaces in the field were thinly mulched with grass to keep tomato fruits clean. Mulch was not readily available. However, unevenly spread mulch decomposes quickly and results in many fruits lying on bare soil, increasing pest and disease incidence. Wilt and blight often affected the local varieties, a reason why vegetable farmers used chemical sprays as frequently as once a week. Wilt has no cure apart from resistant varieties. Pesticides or insecticides (e.g. Super Ambush) were mixed in the same spray pump with fungicides (e.g. Mancozeb and Dithane M45). Commercially oriented tomato growers preferred growing tomatoes in the dry season

Table 7: MAK-ISPUH project interventions to improve pesticide use and handling.

Problem in field	Intervention by project	Remarks
abuse of chemicals	• sensitization and education about safe use and handling of pesticides	• types of pesticides and application rates and methods • safe handling of chemicals • effects of pesticides to human health
	• provision of protective gear (face mask for mouth, nose and eyes, gumboots for the legs and feet, gloves for the hands and relatively strong overall for the rest of the body)	• use of protective gear
susceptible crop varieties	• experiment on improved varieties	• manglobe and heinz v. local varieties, clean and certified seed, sensitization on causes and management of diseases
agronomic issues	• experiment on nursery bed v. direct planting • use of manure to step up soil fertility • pruning of branches	• identification and use of strong and healthy seedlings • 3 branches left per plant (one erect and one on each side of the plant) to increase number and size of fruits
	• staking	• support climbing of plant vertically to minimize incidence of blight due to close interaction of branches with the soil as they creep on the ground
	• spacing	• 90 by 45 cm for tomatoes • 60 by 60 cm & 60 by 45 cm for cabbage

because of assured high prices due to seasonal scarcity, and often used swampy areas with moist soils during the dry season.

An entire complex of farming system problems is bundled together in this case. The FFS sought to work with farmers to try and develop both better agronomic practices and also to reduce pesticide use and abuse. The range of interventions is shown in Table 7, and included variety introductions, agronomic changes intended to reduce dependence on chemical spraying, and a number of safety measures. But somehow, the package failed to cohere. Farmers found valid objections to most aspects, which the FFS approach seemed powerless to overcome. Introduction of certified improved tomato seed varieties was intended to minimize incidence of disease due to resistance. Improved varieties, though still susceptible, are less prone to blight than local varieties, a disease which attacks the crop after fruiting. But these disease-resistant varieties tend to be more demanding of fertilizer. One such variety is "Heinz". Despite its larger fruit, farmers reject the extra costs of fertilization it requires, since it commands only the same price as other tomatoes (200 Uganda Shillings) on local markets.

Improved agronomic practices, such as nursery bed preparation, staking, mulching, pruning, spacing and use of improved varieties, result in higher yields (30-40 fruits per plant) with bigger fruits. But not all farmers were willing to accept the expenses involved. Use of recommended spacing (60x45 cm) and staking saves space and minimizes blight, thus reducing spray costs. It also protects fruits from rotting. Monitoring and rouging also helps reduce the need to spray. But although clean and healthy looking tomatoes result, staking is very demanding of time, given that the farmer has to look for pegs and drive them into the soil, as well as keep directing the climbing plant round the numerous pegs (one per heap). Farmers with smaller fields tried it but those with bigger fields found it very labour demanding and expensive. Some farmers, for instance in Mbarara, innovatively extrapolated the stakes to a mesh-like structure. Although this structure supported loaded tomato stems from falling off the pegs, it required more pegs than staking. Farmers are less than keen to take on these new labour burdens when the rewards are uncertain or unproven.

Use of protective clothing was not taken seriously. While spraying, farmers feel more comfortable with less clothing on their bodies due to heat: men often sprayed half naked without even shirts. Some used their normal clothes and did not change or remove them after spraying. Many farmers are poor, and have few changes of clean, decent clothing. It seems to them a waste to have to change clothing after spraying. Protective clothing was used more to teach others about how and why to use it than it was used by those demonstrating it to protect themselves! In other words, the FFS became an opportunity for rhetorical display rather than communication of practically effective knowledge. Women were the majority in the FFS but many felt uncomfortable with overalls as protective clothing, and there was no scope to discuss the issue. Traditionally, it is against local culture for a woman to dress like a man and yet overalls (i.e. protective clothing) are seen as "male" dress by virtue of their trousered design.

Farmers had a conviction that chemicals were dangerous when taken through the mouth, but the skin, nostril, eyes and other bodily openings were not considered important avenues for chemical entry. Furthermore, farmers spray randomly, and as convenient. This means at times they are working against the direction of the wind, and thus inhale chemicals swept back by wind. Very minimal precautions are taken while handling chemicals. Mixing chemicals in a spray pump was by hand. There was little knowledge that frequent exposure to pesticides and accumulation through the mouth, skin, eyes and nostrils might lead to acute and chronic effects over time. Farmers were told about the danger of unsafe pesticide handling and safe pesticide practice demonstrated to them. Though human health is very important, emphasis was not put on safe pesticide practices (emphasis was on pest identification and control). Farmers see the mouth as the only vulnerable orifice, and they could not relate with the teaching on safe pesticide practices. No one will use hot clothing if there is no understanding of the nature and long term effects. The project did not take further steps to communicate the longer term dangers associated with exposure to pesticides and to make the protective gear more feasible to wear/use. Inability to address this issue implied failure of the project to cope with a major participatory challenge of analysing the local practices.

Buying chemicals in smaller (retail) quantities and use of rates perceived to be appropriate would have been an important objective for FFS in relation to growing tomatoes. Recommended rates are rarely followed due to both language and computational difficulties. Measurements are given in hectares - a unit not understood locally, and manufacturer's instructions on the containers are in English. Farmers are at a loss as to how to calculate the minimum necessary does for small areas. Moreover, different farmers have access to spray pumps of different capacities (ranging from 5 litres to 25 litres). Some (though very few) improvise with basins and use leaves to sprinkle pesticides in the field. Information on the right dosage was limited to the pesticides used in the experiments and for a 20 litre capacity pump. Farmers were just told the right dose and not taken through the process of how to make calculations for the right dosage, even for other pesticides that farmers were likely or used in their home fields. Although farmers were often advised to get the right pesticide application rates from the agro-stockists from whom they bought, most agro-stockists were not technical enough to guide farmers in dosage (they are business men/women) were ignorant of the right application rates. Some gave higher dosage and others gave a lower dosage. Helping farmers know how to make these calculations would contribute to adequate pesticide use and minimum abuse of pesticides by farmers.

One important point to emerge from the MAK-SPUH case is the importance to tomato farmers of local market signals. Tomatoes with visible spots of powder (DithaneM45) were in higher demand in the market than clean fruits without milky/powdery chemical residues. Consumers had a belief that the prolonged shelf life effect of DithaneM45 reduced perishability, and thus actively preferred tomatoes showing evidence of recent spraying. This even encouraged spraying of tomato fields close to harvesting, with the risk that fruits might be eaten fresh/raw without washing, especially by the harvesters themselves. This instance suggests farmers defend certain local (and dangerous) practices not out of some misguided technical consideration of their own but because they are responding to (perverse) market preference. Analysis of the MAK-SPUH case thus suggests the need to enlarge the scope of FFS. It is important to understand how new technology fits not just within farmers' production environments but also within the existing local market systems. Farmers', especially the more commercially oriented, will readily take up a new technology if its use leads to increased income: increased yields that respond to no market signal offer no motivation. In other words, producers listen more to what buyers want than what scientists say, implying that FFS needs to engage buyers and consumers perhaps as much as producers in regard to issues of safe pesticide use.

4.2.3 A2N-ISPI project technology intervention - soil fertility improvement

Improving upon soil fertility to ensure increased agricultural production and productivity is the critical issue for A2N-ISPI. Soil fertility in Uganda is said to be generally on the decline (Zake 1993), with eastern Uganda being the region most badly affected mainly due to the sandy nature of the soil (Bekunda *et al.,* 1997; and Wortman and Kayizzi, 1998). Farmers do little to revive, improve and/or maintain soil fertility. In Tororo, a majority of farmers did not carry out soil conservation practices, yet annual crops, generally cultivated continuously, exploit soil fertility unsustainably. A recent diagnostic survey on farming system and soil management in Tororo (Delve *et al.,* 2003) revealed that more than 80% of farmers used neither inorganic nor organic fertilizers. A very few did use inorganic fertilizers in commercial fields of rice, maize, groundnuts and onions, but not continuously, due to expense and difficulty of access. The belief among farmers that inorganic fertilizers spoil soil contributes to them not bothering to seek out and use the fertilizers available.

Using Serenut II groundnuts and Longe 5 maize (improved varieties) as study crops, experiments on different organic and inorganic fertilizers (UREA, Di-ammonium phosphate and single super phosphate) were established. FYM, compost and green manure formed the organic fertilizers. The green manures included *Dolichos lablab, Mucuna pruriens var. utilis, Tithonia diversifolia, Lantana camara, Canivalia ensiformis* and *Crotalaria ochroleuca*. Farm Yard Manure (FYM), as an organic fertilizer, was used only in a few small back yard kitchen vegetable gardens because of the belief that it was for the rich. "People keep telling us that use of FYM is for people who have livestock (cattle) for the dung, can afford a wheel-barrow to ferry it to the field and are able to pay labourers because it is a tiresome and demanding job..." Farmers did not go commit themselves to go through the technicalities of preparing FYM, where a mixture of animal waste (mainly cow dung), dry and fresh crop residues, and other materials are put in a well-dug pit to rot. In the FFS context, FYM refers basically to nothing more than soil collected from around kraals. Use of green manure was negligible.

Farmer practices in relation to soil fertility project interventions
In spite of the exposure in FFS to the above interventions, farmers did not take up soil improvement technologies, mainly due to lack of immediate benefit, time, and labour, and cost implications, much as has been described for MAK-IPM project interventions as described in Section 4.2.1. Yield improvement is a long-term gain, but the green manure crop itself has no immediate beneficial use, e.g. as food. *Tithonia diversifolia* and *Lantana camara* require to be cut from the bush or by the roadside where they grow wild, and then ferried to the field, chopped and ploughed back into the soil. *Lantana camara* is thorny and pricks the handler while *Tithonia diversifolia* leaves a bitter taste on the hands. Furthermore, people with cattle are very few and therefore it is a problem to obtain FYM, despite FFS experiments

with FYM (on both maize and groundnuts) revealing better crop performance and yields. Some studies conducted in eastern Uganda (Olupot *et al.*, 2004) suggest that even where the high quantities of manure required are available (2.5t ha^{-1}) quality is not adequate and requires combination with mineral nitrogen and phosphorous fertilizers. Most farmers cannot afford agro-chemicals, however, and in addition lack kraal manure - having very few who have cattle. Use of phosphorus fertilizers to boost performance of *Mucuna pruriens var. utilis* was a discouragement to farmers besides wasting land for a whole season under *Mucuna pruriens var. utilis* that had to be treated like any other annual crop! *Crotalaria ochroleuca* was readily infected by pests and needed spraying. Hence, it was not worthy for farmers to invest in the technologies.

Analysis of the A2N-ISPI case reveals that resource-poor farmers consider use of fertilizers as an activity for the rich. That wealth differences have an effect on social solidarity comes out clear through the idea that fertilizer spoils the soil. Building on existing practice requires a comprehensive understanding of the hidden rationale regarding what farmers do and using the rationale to design feasible solutions to the existing problem. Technologies promoted in this case needed to put into consideration the availability, accessibility, and affordability of raw materials required. Because soil improvement with limited means is a complicated process, farmers cannot simply prioritize soil improvement once a project team points out potential benefits.

4.2.4 CIP-IPPHM project technology intervention - sweet potato productivity

The critical issue the CIP-IPPHM case attempted to address is to increase productivity of sweet potatoes, an important food and cash crop in Teso. Production and post harvest handling were the areas of focus. In Katine and Abuket (sub-counties in Soroti), where CIP-IPPHM implemented FFS activities, farmers grew cassava, sorghum and sweet potatoes as the food crops in that order of importance. Women were more involved in these crops than men. Of the cash crops, millet was the most important, followed by groundnuts, cowpeas and sesame respectively. Farmers explained that there was only a minor involvement in commercial sweet potato production, mainly due to the fact that the white-fleshed roots of local types were not liked on the market. "*Buyers in Kampala prefer the orange and yellow fleshed sweet potatoes...*" Farmers talk of the city (Kampala) as their market site, yet by their own confession few have gone there to trade. Sweet potato is mainly a subsistence crop. One wonders if availability of orange fleshed sweet potato would lead to commercial sweet potato production in an area where even the local white fleshed type is not produced for local sale. Farmers showed interest, given their alleged marketing in the capital, a claim that probably helped them to get access to orange sweet potato vines. One of the roles that FFS might play would be to encourage farmers to be proactive in seeking and accessing adequate market information (about, for example, prices, qualities required by market and alternative market places where local produce might be in demand). Unfortunately, FFS in Uganda suffers from a production bias. Market information is a key issue that tends not to be addressed.

Farmer practices to ensure sweet potato vine availability

Sweet potato production in Soroti is a second season activity. The peak of the second rains is experienced between May and August. Millet was the main crop following sweet potato in rotation. This second rainy season is followed by a long dry spell, often stretching from December to April. During the dry season almost all sweet potato vines in the field become scorched and dried under the hot sun. In search of pasture, livestock feed on any remaining vines that can survive the sun. Scarcity of vines (Bashasha *et al.*, 1995) results in chronic late planting of sweet potato, year after year. However, since it is a major food crop, farmers have developed mechanisms and strategies to manage vine availability to ensure availability of food. These strategies include:

a. Plot reservation and use of tubers: farmers reserve some sweet potato plots specifically for production of vines. They leave these plots or gardens intact - never harvesting any tubers or vines. The idea is that after the vegetation has dried, the tubers in the ground will still remain potential producers of vines. As soon as the rains began, tubers in the reserved plot (mounds) begin sprouting, and yield vines that can be later harvested for planting in newly prepared fields. In a rush to plant early, to catch up with the season and to reduce the risk that other people will take the scarce vines, farmers harvest these sprouts as soon as they appear, cutting the vine just at the base where it emerges from the soil. The tuber is commonly infested with weevils, and as a result farmers perforce use infested materials. This perpetuates sweet potato weevil season after season.

b. Establishment of sweet potato gardens under a large shade tree. Farmers mention trees that are specific to this purpose, where vines perform better. The mango tree is not recommended because vines under a mango tree are reportedly more readily attacked by insect pests. Given that the area is one of the leading producers of mangoes in Uganda, it would be worthwhile to understand the logic behind this claim that mango trees foster pest damage. FFS did not investigate this issue.

c. Planting sweet potato vine gardens in swamps during the dry season (December-April). Soils in valley bottoms remain moist in the dry season and are more fertile. During the dry months normal farm work ceases. The dry land soils are too hard to be worked and the sun is too hot.

Some women raise vines in small gardens near their homesteads where they are able to water daily and drive away livestock from feeding and destroying the planting material. During the dry spell, availability of water is a problem. The women pour water earlier used for washing or food preparation over the sweet potatoes to keep them moist. Those who manage to raise some vines early enough in the season can sell them at a good price. Depending on availability, month and variety, a sack of sweet potato vines goes for US$3.5 - 5.5 (7000-10000 Uganda Shillings), while a bundle (handful) sells for US$0.2 - 0.5 (500-1000 Uganda Shillings).

Farmers' sweet potato weevil management: Sweet potato weevil (*Cylas formicariuse elegantulus*) is an insect pest in the group of beetles that damages the roots, stem and leaves of the sweet potato crop in the field. The larvae tunnel through stems and roots causing brown lesions in the inside and small dark holes or perforations scattered on the surface of the sweet potatoes. Infected sweet potatoes have a bitter taste and a characteristic smell - un fit for use. Farmers have stayed with this pest and have varied knowledge about how to manage this pest. There is no chemical control as yet for the weevil. The control measure is by destroying infested plants, removing volunteer sweet potatoes and other related weeds that may host the pest. The farming practice of using volunteer crops as source of scarce planting materials, however, has trapped sweet potato farmers in the cycle of the pest. Farmers have varied knowledge about the weevils and their control.

Some farmers admit they do not know what to do and just leave everything to nature. They have lived with the problem and have all but given up. They believe weevils come from the soil. Some said "the weevils come from the caterpillars that we see!" Weevils migrate from infested sweet potatoes and related host plants like Morning Glory to new sweet potato fields. Maintaining distance between new and old sweet potato fields was not a pest management practice with which farmers were familiar. They see no reason not to establish a new sweet potato field adjacent to a previous or existing one (Ebregt *et al.*, 2004). Sometimes the same field carries a sweet potato crop for two or more consecutive years. This practice certainly contributes to build-up of the weevil. Early and complete harvesting, instead of piece meal harvesting, was among the common practices used by some farmers to try and minimize the effects of the weevil on the quality and quantity of the yield. Harvested sweet potato roots were then sliced manually and sun-dried for future consumption.

Local sweet potato varieties: In evaluating varieties convenience of cooking (time and fuel) and harvesting sometimes outweigh issues such as resistance to pests. Among the local sweet potato varieties grown in Soroti (*Araka, Obwana Alwala, Muyambi, Osukut*) two (*Obwana Alwala and Muyambi*) were said to be tolerant of both weevil and drought. However, farmers did not like them because of their inconvenient characteristics. Development of root tubers deeper in the ground and away from the mounds made harvesting more difficult. Thicker skins made peeling tiresome. Too much sap was an inconvenience during harvesting and when peeling in preparation to cook. The tougher vines could not be easily broken in the process of harvesting them for planting using bare hands (i.e. they necessitated a knife) and tougher tubers took a relatively longer cooking time, therefore requiring more fuel. Marketing was noted to be a major problem especially in the period from December to February when every one harvested sweet potatoes and therefore there were no local buyers locally.

Project interventions to improve sweet potato productivity

Confirmation of sweet potato as one of the most important crops of the area was an opportunity for the project. Sweet potato-related problems mentioned by farmers included vines scarcity, pests and diseases, and lack of marketable varieties. Farmers experienced shortage of vines at planting time, swellings at the base of the vines, rotting of tubers, bitter taste, pimple like structures on the skins (scab), millipedes damage, lack of markets, failure of the roots to fill into tubers, twinning of vines, hairiness of the vegetation associated with lack of tuber formation (probably viral disease) and storage pests. Of the mentioned problems, lack of planting material, weevils and marketing were ranked most important. These served as good entry points for the project and actually fitted in well with what the project already planned to do will be seen below.

Scarcity of planting materials: For the problem of lack of or scarce planting material, two things were done. One, introduction of improved (orange fleshed) sweet potato varieties that were presumably resistant to the weevil and with a relatively shorter maturity period and two, rapid vine multiplication technique. Rapid vine multiplication method was introduced to the farmers to learn how best to multiply vines in the shortest time possible. A small plot was prepared at one farmer's place and treated with compost: the more fertile the site, the faster and better the results. After preparing the bed, clean vines (of the desired variety) were cut into pieces of about the length of the longest finger. Half the trial was planted with cut pieces (of 3 nodes) planted in the prepared bed vertically with one node buried in the ground/soil. The other half was planted horizontally allowing shoots to sprout from the buried node areas. This was followed with frequent watering (on at least a daily basis). In this way, farmers raised their own vines. The sprouted shoots were ready for harvest as planting material after attaining a length more than about 30 cm (estimated visually). Of the two rapid vine multiplication methods used, the vertical method required less labour and shoots sprouted faster than the horizontal method. It was also closer to the traditional method used in planting sweet potato vines.

Normal vine length for farmers in Soroti is/was about 20 cm, shorter than the recommended length of 25-30cm. Experiments on vine length were also set up with treatments of difference in lengths: 40cm, 30cm and 20 cm (farmers' length). Although longer vines established faster and established roots deeper in the mound where weevil infestation was lower (about 5% as compared to 20% for shallow planted vines, on average), farmers preferred their (short) vine length. Farmers' vine length was not significantly different from the recommended 25-30cm length. Preference for the normal 20 cm vine length was mainly attributed to scarcity of vines as well as in relation to mound size.

Farmers in Soroti heap small sized mounds to ensure efficient use of atmospheric moisture during dry periods. Small mounds lose moisture very fast during hot periods of the day (i.e. late afternoons) but were more efficient at absorbing the moisture present as dew in the mornings and evenings. This protected the developing sweet potato roots from rotting. According to the farmers, although big sized mounds take longer to dry or lose moisture, it was more difficult for moisture to penetrate through the big mounds to the roots especially during the dry periods. The small sized mounds implied pushing the longer vines to a greater sub-soil depth, which was hard and therefore gave poor vine establishment. Although small mounds were easily washed away by rain, given the sandy nature of the soils, farmers earthed up the mounds as the plants became established. Farmers were encouraged to mulch in spaces between mounds and/or ridges to conserve moisture in the sweet potato fields. However, it was not very easy to get mulching material, especially for bigger fields.

Control of sweet potato weevils: Use of clean planting material, planting methods, minimum distance between old and new sweet potato garden and right agronomic methods were tested to help farmers realize various ways of managing the weevil problem. Healthy looking vines cut above the base had less chance of being weevil infected as compared to unhealthy looking vines cut at ground level. Vine multiplication and use of clean planting material addressed the most pressing need of vine availability and weevil management without changing farming pattern/style.

Sweet potatoes are planted in three different ways depending on the country. Although mounds are becoming more common in Kenya over time, cultivation of sweet potato using flat land is the more traditional planting approach in western Kenya, while ridges and mounds are the traditional methods in Tanzania and Uganda respectively. Use of mounds and ridges were set up as experiments to decide which was better in minimizing weevil infestation. Farmers in Uganda heaped the mounds almost one on the other, a negligible distance apart. Two common planting methods (mounds and ridges) were experimented using the same variety. The recommended spacing between mounds under the project was one metre (1m). This served to minimize chances of weevils and perhaps other pests easily crossing from one heap/mound to another. To most farmers, the 1m space was not welcome. It was a waste of land, where they wanted to maximize use of limited resources for maximum output. Some farmer diplomatically resisted new technologies he perceived to be inappropriate in his context (Box 8).

Inferior varieties: The white and cream-fleshed local sweet potato varieties were susceptible to weevils and less preferred on market. The project introduced six new orange-fleshed varieties that were evaluated for their performance in the farmers' contexts. *Ejumula* and *Kakamega* (as named by farmers) were preferred because of their yield and reduced susceptibility to weevil attack. In the interest of saving space, time and labour, each plot had 24 mounds of farmers' size

---

Box 8: A diplomatic way of resisting a project.

After demonstrating to the farmers of Omodoi and Abarigentie FFS, who had gathered for the preliminary meeting and exposure on how layout of trials could be done with a spacing of about one metre to separate the different plots, Mr. Ocen who had offered to be the one to have an individual trial on his land (the very one who had offered his ploughed field for group activities) abandoned what we had agreed, heaped mounds in a different way (continuously with no space) and planted his own local vines that we did not know. While explaining about the trials he inquired who would heap the mounds for his individual field. "Then the rest learn from my sweat?" he asked with concern. He even mentioned that leaving 1m between plots was a waste of land. On delivering planting materials for establishing the trials, none of the family members was at home to receive the vines. Looking at the field upon which we had agreed to have trials, there were vines already planted. We were stuck. "He went somewhere far" said the kids who kept a distance from us. Eventually we gave the vines to some lady who looked willing to do it and left, after guiding her on how to lay the experiments.

---

and each mound planted with two vines - because of vine scarcity. In addition to collectively managed experiments at (4) FFS sites, one volunteer farmer in each FFS group planted the same experiment at his home for replication purposes. However, individual farmers were tempted to harvest some vines and roots out of curiosity, which affected reliability of the outcomes of the experiment. Others were just selfish, as one farmer made clear when declining to host an experiment after realizing that it was for the learning benefit of all group members (Box 9). To this farmer, hosting the group experiment meant sharing the resulting vines and roots, which he was not comfortable with. Variety evaluation became an FFS activity in 2004, after individual farmers failed to manage the experiments at their farms the previous year. Farmers were not expected to harvest anything, from the variety trial plots before the final collective evaluations had been made. This was respected in 2004. Replications allowed farmers in FFS make more informed decisions on what variety to take up.

Despite the attractive orange-fleshed colouring, higher yield, tolerance to drought, early maturity and reduction in fibrous characteristics, some farmers and consumers preferred local varieties to the new varieties. Local varieties were sweeter and had higher starch content than the new varieties. This made them firmer to the taste, a characteristic liked especially by adults. They were also more resistant to diseases (Mukasa, 2003) and thus performed better than official varieties (Abidin *et al.*, 2005). Softness or watery nature of the tuber when cooked, less taste, an undesirable characteristic smell, especially when harvested before the recommended maturity period, and susceptibility to sweet potato viral disease (SPVD) were among the general reasons why improved varieties were seen as less preferable than local varieties.

In Bungoma (Kenya) some farmers also preferred a local sweet potato variety (called Bungoma) to the improved orange fleshed varieties *Ejumula* and Kakamega (scientifically coded as SPK004) because of taste (sweeter), more ready local market (traders come to the villages looking for it), and better yield. They even asked the breeders (informally) whether they could improve the local variety, e.g. by including the orange flesh colour. Farmers in Kenya expressed insecurity in taking up the improved varieties. This was reflected in one farmer's words explaining why she was not comfortable with improved varieties "We are used to our local type and know it as reliable as a food source. We fear to lose it in favour of the new types whose reliability we are not even sure of…" According to farmers, improved varieties lose viability over time, a reason why they keep disappearing. "If you leave your culture, you become a slave…" loosely translated from the Swahili saying "*muacha mila ni mtumwa*". This lack of confidence in new technology decreases chances of farmers taking innovations up. This shows the importance of existing farming practice as an outcome of a long-term learning process. Relating to such a learning context with new technologies (from another context) can only succeed when the learning process becomes a common practice and does not turn into a one-way instruction mode.

Market value of sweet potatoes: For the issue of sweet potato marketing, the project introduced record keeping where farmers use costs and incomes to identify profitability of sweet potato production venture, and encourage farmers to identify and concentrate (supply) on varieties liked on market (demand), and sweet potato processing into other forms that could be used and sold for income. Two sweet potato chipping (processing) machines were provided to FFS groups (courtesy of CIP, in partnership with NARO) in an effort to add value to the sweet potato after harvesting. As a result, a sweet potato processors association were formed and made linkages with potential milling companies to buy the chips. At a local level, the chips were dried and ground into flour used to developing various products sold locally, especially to school pupils. The flour was used as an ingredient in preparing porridge and pastries (chapatti, cakes, doughnuts, crisps). Other products from the orange fleshed sweet potato include juice and soap. However, these are still at the local level. There is yet a step to take in packaging, formulation and quality assurance. Keeping farmer records was still a weak point on the farmers' side. This implied that they were not yet very commercially oriented. But even so the machines helped sweet potato producers to realize the commercial value of the crop beyond home consumption. Because of its soft nature and orange flesh colour (with high vitamin A content) processors preferred the improved varieties to the local varieties.

The sweet potato processor groups mentioned a number of challenges they face. These include: (a) inadequate quantities of tubers for processing due to competition with the market for fresh produce (b) Unreliability of buyers of the chips, (c) inadequate storage facilities, and (d) the rather low quality of the chips, which sometimes fetched less money than expected (e) "selfishness of fellow processors" (competition), especially in Kenya (including monopoly of supply to some industrial users, such as a plant in Kirinyaga and (f) poor location (being based in an upcountry district makes it difficult for the processing association in Soroti-Uganda to hunt down potential buyers in urban areas. However, the chairman of the association seizes all opportunities (informal and formal) to promote the association and its activities.

Technology interventions selected by researchers as suitable solutions to farmers' problems do not necessarily fit within farmers' own framework of interests and priorities. Incompatibility of improved practices (vine length, planting method, and isolation distance) with traditional practices of sweet potato production, especially in terms of labour required, minimises farmers' interest in new practices. New interventions need to be realistic in relation to local contexts and resources if they are to be useful to the communities where they are introduced. Farmers are often perfectly rational, and abandon technologies that seem to demand more than they can afford, especially when they discover local practices to be more "user friendly" than new ones. One of the hopes for FFS is that it could become a methodology for undertaking a kind of participatory feasibility study, throwing light on what farmers do and why, so as to establish a better match between local needs and improved methods. But researchers' priorities and mind sets tend to undermine this aim. As with cowpeas and groundnuts in the case of MAK-IPM, farmers in CIP-IPPHM also preferred the local varieties to improved orange fleshed sweet potatoes because of factors not fully allowed for by researchers (e.g. taste) What researchers take as very important is often different from what farmers take as important. Researchers tend to focus on yield and resistance, whereas farmers (reflecting a subsistence orientation) are first conscious of taste and effort (labour efficiency). But the FFS presently discussed has generated some useful results about what farmers might consider ideal in varieties combining new features (such as orange flesh) and good attributes (e.g. firmness and sweetness) associated with local varieties.

Clinging to traditional practices: implications for new practices

In spite of vulnerability to erosion and higher weevil infestation due to exposure of roots to rain wash, most sweet potato farmers in Soroti preferred using their traditional mounds to ridges. Mound sizes suit different rainfall regimes. The sandy soil is vulnerable to rain wash as well as desiccation due to drought (Aniku, 2001). Findings by Ebregt *et al.* (2005) seem to suggest that sandy soils enhance invasion by and activity of sweet potato weevils. During high rainfall events, large mounds are preferable: rains do not wash them down completely, hence reducing the chances of exposing root tubers to rotting in very wet conditions. During dry periods, smaller mounds are preferable since they enhance infiltration of rain water. Making ridges demands more time and labour especially when using hand tools, yet access to oxen was very limited and expensive. The size of the ridge depends on the number of times the oxen were driven around the same ridge: the more the times, the bigger the ridge. Use of oxen simplified work, especially in Soroti where the farming system involves use of ox-ploughs, but this was mainly accessible only to commercial producers who can afford investment or hire. The grave-like appearance of ridges made some farmers feel uncomfortable in psychological terms. Farmers observed that mole-rats (in effect a large burrowing rodent) could easily clear a whole ridge were they attack a field.

The idea of ridging was mainly taken up by commercial sweet potato vine producers and processors for quick canopy establishment, minimal weevil infestation, and adequate tuber size (as required for processing). From good quality (clean with no weevils) vines farmers (in Abuket FFS) generated income through contracts with other farmers (individuals and groups) and organizations within and outside the village and district. The organizations to which they sold material included the Agricultural Productivity Enhancement Program (APEP) and World Vision-Uganda. Mounds produced undesirably large-sized sweet potatoes. Besides being demanding of time and labour, chipping the big sized sweet potatoes also caused problems, since larger bits remained un-chopped (a waste, in fact). In an informal discussion with farmers and sweet potato processors from Kenya and Uganda during one stakeholder workshop in Busia (Kenya), views about plant population and root size varied.

With the recommended spacing of 90-100 cm between ridges and 30 cm from vine to vine one could have one or two lines per ridge. According to one (male) processor and farmer from Uganda, planting vines in one line produced better yields, but with larger tubers not desired by the processors. But a female farmer/processor from Kakamega found that two lines on one ridge gave higher yield and a desirable size of sweet potato for processing, as well as tubers with higher dry matter content. Women have more experience in potato production compared to men, given their traditional engagement with the crop for home consumption. Difference in ridge width and soil fertility might explain differences in tuber size. In the Ugandan context, the width in question was about 50-60 cm, while in the Kenyan context it was about 100 cm. Soils in Soroti - specifically in Abuket - may have been relatively more fertile compared to soils in Kakamega. Other advantages attached to use of ridges include convenience in moving through the field and conservation of moisture and soil.

In the traditional Uganda manner, two to four vines were driven/pushed/planted in one mound at a single point (via the peak of the mound), leaving only the top leaves hanging out of the soil. The number of vines per mound depended on vine availability, soil fertility, survival expectation, production goal, and maturity period. The newly introduced method, however, encouraged use of three vines, each pushed in the mound/soil one at a time and via one side of the mound - i.e. making a triangle. Besides assisting formation of a canopy around the mound in a shorter time, the 3-vine method minimized the number of vines per mound, hence it was realistic in a situation of scarce planting material. Early canopy formation by creeping vines smothered weeds and provided a micro climate unfavourable for migration of weevils and other pests from mound to mound. However, given the planting method used, three vines planted singly placed farmers' hands at greater risk of being hurt by stones or sharp sticks while planting the vines and had implications for time and energy spent per mound/field. This was in effect to ask women to use more time and labour in planting sweet potatoes, an investment not all could or would make.

Farmers did not object to the lack of distance between their new and old sweet potato fields, yet according to researchers, short distances promote pest migration from old to new field. The recommended isolation distance between old and new field was about 100 metres. During a sweet potato season almost all farmers will plant the crop anywhere and everywhere, given opportunity, and it becomes difficult even to find a location in which such a distance can be maintained between one sweet potato field and another. This particular experiment was unrealistic in relation to the farmers' context. Introduction of 100 m distance as a pest control strategy assumed that all farmers regarded sweet potato as very important, and that all farmers perceived the weevil as a major threat to their food or income security. In practice, sweet potato was only one of the crops through which farmers had diversified livelihoods, and not necessarily their main priority, so long as they could grow some. If the 100 m had been enforced, it would result in some farmers failing to grow the crop because there simply were insufficient suitable sites to go round. In short, this technology was not sensitive to realities on the ground.

Shorter distance, with taller crops/plants serving as barriers to migration of weevils and flight of white flies transmitting sweet potato viral disease (SPVD), was introduced as another experiment. The experiment was set up using 5 metres between old and new fields (the average of the set of shortest distances between farmers' sweet potato fields). One space was left bare and another was planted with a local cereal crop (millet, maize or sorghum). One FFS group managing to conduct this experiment found no difference in weevil infestation between bare and cereal-barrier plots. No results were generated on white files and viral infestation. This experiment in isolation distances and use of guard/barrier crops was meant to be repeated, but farmers in the FFS were not keen. It was a bother to plant another crop/cereal during the second rains. Cereals are planted during first rains when pest pressures are lower. Legumes and root tubers are mainly planted in the second season. Planting cereals again in the second season implied to many farmers a waste of scarce time, energy and seed, especially given the higher pest pressure on cereals and unfavourable climate.

4.3 Positioning FFS interventions in existing farming systems

The present section will discuss the problematic issue of how FFS interventions are positioned within local contexts. Do they really address local priority issues? How in general can research identify such priorities, and is there ever a single set of such priorities, or as many lists as there are different local interests groups? How is participation to address this issue? Here it makes sense to pay a little more attention to the general setting within which FFS in Uganda (as reflected in the cases addressed in this study) finds itself.

Farmers in Soroti and Busia grow the same crops and are involved in the same activities. The major crops ranged from cassava, sweet potatoes, cotton, groundnuts, peas, beans, rice, millet, sorghum, maize and sesame. Whereas cassava is the main food crop in Busia, millet and sorghum are the major food crops in Soroti. Both districts have two farming seasons, based on a bimodal pattern of rainfall with peaks around April-May for the first season and around August-September for the second season. Unreliability and uneven distribution of rainfall pattern, together with the soil type, dictates patterns of cultivation of and dependence on annual crops. Presence of perennial crops - especially coffee and bananas - in some parts in Busia suggests a cooler and moister micro-climate. Livestock (oxen) are a major component of the agricultural system in Soroti. Soroti has a population estimated at 446,300 in 2006, and covers an area of 3,377.7 km^2, of which 1,809.2 km^2 is farm land supporting a rural population of 355,926. With a total land area of about 759.4 sq.km and population of 225,008, farmland in Busia district comprises 528.3 km^2 and rural population is 205,518 (see UBOS, 2006). People in the two districts engage in similar income-generating activities, but with variations in intensity. Agriculture, fishing and fish mongering, petty trade, carpentry and blacksmithing, brick making and civil service provision are among the list of income generating activities in which rural populations are involved. Many of these livelihood activities are combined with some amount of farming. Women are largely involved in food crop production, and therefore control food security, while men are mainly engaged in commercial crops, hence are expected to provide most if not all the financial needs of the home.

Presence of food on the table is the starting point of domestic life. When there is food, other problems can be discussed and solved. But in absence of food the only problem that everyone is likely to talk about and seek amelioration is lack of food. Other problems tend to be backgrounded where food is a problem. FFS interventions help farming communities produce more food. However, based on the evidence of the FFS interventions reviewed here, food does not seem to be a critical issue. An implication is that people have food and now need other things as well. FFS technologies made contributions towards food security through addressing principal causes of low crop yields, but we might ask whether in broader regional context this priority was well chosen.

4.3.1 Farmers' perceptions of priority issues

Understanding what goes on in a community is somewhat easier an analytical task than to focus on community needs, which necessitates stratifying a community to identify what each interest group or class perceives to be its need. Results vary across the different strata and between strata and the group as a whole, which shows the difficulty of achieving answers to common problems. Variation in importance of a given issue as perceived by different categories of people in a community (Table 8) revealed the complexity of entering the mine-field of perceived needs. For example, in Busia where the A2N-ISPI project operated, lack of income generating activities was perceived as a major problem by many. In Soroti, where CIP-IPPHM operated, malaria and HIV were perceived as major problems. These perceptions, influenced by a number of factors, vary from context to context, and make it difficult to establish a fit between technologies or projects and local situations. Though non-FFS farmers perceived yield issues as very important it did not seem a priority problem as reflected in the general position in the ranking. For instance low yields ranked fourth in Sihubira village (Lunyo sub-county, Busia) district) and fifth in Abuket village (Kyere sub-county, Soroti districts). Ranking low

Table 8: Variation in farmers' perceptions of priority issues in Busia (S) and Soroti (S).

Issue/problem mentioned/listed and later ranked	FFS alumni ranking				non-FFS alumni ranking				All	
	Women		Men		Women		Men		Collective ranking	
	S	B	S	B	S	B	S	B	S	B
Marketing of produce	1	5	2	4			1	1	3	3
Low crop yields (reasons included - in descending order of importance - pests and diseases,* poor soils, especially in busia, scarce and expensive improved seed, lack of ox-plough, labour problems and local tools, unreliable weather (drought & hailstones), lack of storage, insecurity, due to rebel activities))	3		4	2	1	1	2	3	5	4
Malaria and hiv/aids (diseases and unreliable distant health centres)	2		1	3	4		3	4	1	2
Lack of togetherness in community			3	1						
Lack of income generating activities	2	1	5	5	2	3	4	2	4	1
Transport problems	4		6		4					
Lack of safe water	3		7		3		5		2	5
No opportunity for women to contribute to decisions about using money in households	5	4			5	2				

* insect pests like aphids are more of a problem in Soroti as opposed to vertebrate pests like mole rats and monkeys in Busia. S is an abbreviation for Soroti district and B for Busia district.

crop yields in Soroti, for instance, as a low priority may have reflected a village perception that the project focused on this too much already.

Focus on low priority issues in a given community risks losing farmers' interest in FFS. In order to maintain interest in the whole project, attempts were then made to include items of greater interest as special topics. These included HIV/AIDS and its effect on agriculture, reproductive health and family planning, farming as a business (a common topic across all FFS projects), basic financial management skills, gender and agriculture, teamwork and sustainability, sanitation/hygiene, and basic human nutrition. Where facilitators had no capacity to handle these topics competent guest speakers were invited.

Time allocated to special topics was very limited, and competent persons not readily available, so often these topics were ignored. This is to be regretted, since taking on emerging issues that affect the community as they crop up, alongside the targeted technology of the project, is one way to demonstrate a commitment to make FFS fit better and therefore enhance its responsiveness. Shifting FFS meeting times from mornings when farmers attend their home fields to afternoons when farmers are more available was another way to acknowledge farming system constraints. Encouraging inter- and intra-village or regional visits, at individual and group level, was attempted in some (but rare) cases, and this encouraged sharing experiences but also enhanced socialization and networking of farmers.

Pest management and low soil fertility were perceived to be the most important causes of low yields in farming, and these were issues addressed by the FFS interventions here described. But a caveat has to be noted. Whereas it is true at a broad level that pests and diseases were particularly pointed out to be most problematic in Soroti, and poor soils were pointed out as most problematic in Busia, it still has to be shown that they are considered crucial for the specific crops (sweet potato, maize and groundnut) chosen as a focus for FFS in the communities examined. The 'pinch' of low yields due to pests is felt most seriously in major food and cash crops. Even when not actively involved in problem diagnosis, farmers tend to be interested in any opportunity that promises a solution to low yields of their major crops.

4.3.2 Interconnectedness of priority issues raised by farmers

Irrespective of perceived magnitude of importance, all problems are inter-related. Whatever starting point FFS assumes the chosen approach will have wider implications at a farming system level. An approach from the crop yield perspective, for example, will have effects in other areas, such as food quality. It is interesting to try and track some of these interconnections via the FFS process, and specific problem ranking exercises undertaken by farmers in the project communities. As part of research methodology for this study, I engaged farmers in an exercise where they listed and ranked what they perceived as priority issues in their communities. This exercise, aimed at identifying what farmers considered important in their communities, was done with two FFS groups, one in Busia and one in Soroti districts in early December 2005.

Listing and ranking in order of increasing magnitude of issues was first done by each sub group (Table 8) independently. Later, the whole group[21] ranked the issues mentioned, following discussions during the exercise. This brought out a wide nexus of interconnections. Health affected production activities, farm yields, food security, access to market and therefore income. HIV/AIDS and malaria were rampant and resulted in expenditure to save lives, reduced food production, psychological stress and increasing numbers of dependants (orphans, widows and widowers). Farmers then mentioned that there were hardly any medicines for treatment in hospitals, while private clinic treatment was unaffordable. Health services were distant and sometimes lacked medical personnel. This health problem engaged human and financial resources that might otherwise have been used in food production. Continuous cultivation of land to meet increasing needs (food, medical, income, and clothing, among others) contributed to exhaustion of soils, hence low yields. In spite of the meagre harvests realized, middle men exploited farmers by offering very low prices. In Soroti (Kyere), farmers were even taxed to allow their bicycles to enter the market place.

Other problems mentioned included (1) limited supply of expensive labour, while traditional methods are very tiring. Many times this resulted into late planting (more time spent in opening up more land). (2) Alcohol was felt as a problem by some because (allegedly) 50% of people (especially men) were drinking at times they were expected to be working in the gardens/field. Such people were also accused of failing to perform their responsibilities as parents or spouses. Labour is immobilized through drunkenness. (3) Poor leadership - decision makers were reported to be corrupt. "They cheat and are not transparent..., do not do any follow up to ensure that the right thing is done and often politicize everything..." Some said they had no confidence to speak up and tell leaders what was wrong or disliked. There is a problem here in that FFS relies heavily on group discussions, and at times consensus formation silences criticism (Murphy, 1990).

4.3.3 Looking behind ranking of priority issues

Ranking of priority issue in group meetings is influenced by many factors. Ranking is susceptible to the influence of prevailing situations, but may not reflect longer-term concerns. Van Huis and Meerman (1997) indicate that a hierarchy of agro- ecological and socio-economic problems faced by farmers influences how farmers rank their problems. In the present case, competing influences include (1) political situation, (2) crop production and (3) poverty.

Political bias
The ranking exercise just described was carried out at a time when the country was opening up a presidential election campaign (late 2005). People were divided up between different political parties, especially the ruling National Resistance Movement (NRM) and the strongest opposition group (Forum for Democratic Change, [FDC]). Affiliation often was brought out

---

[21] This group consisted of farmers from different groups within the same village and only a very few from neighboring villages. It involved men and women

in meetings as a problem. Farmers talked about the politicization of almost everything, and the 'promotion' of corruption, cheating and deceitful leaders.

Crop production bias

There was a tendency of the community group to mention things that fitted the discipline of the researcher. For instance, when they started writing down their issues, some were biased towards crop production (seed, chemicals, pests etc) based on their understanding of the job of the facilitator. Sometimes they felt the researcher was a government worker, under the Ministry of Agriculture, and therefore best able to offer advice and help crop production issues. This changed after I reminded them to look at issues in a broader perspective. In an attempt to broaden the frame of reference I frequently requested farmers to close their eyes and respond to the question "what do you see as the key things that need attention for a more comfortable life in the village?" The very fact that this then led to a broader range of problems being discussed indicates that FFS projects exert a shaping influence over farmers' rankings, based on cues and signals already provided about what the organisers might see as preferred content.

Poverty bias

Lack of money is one of the most pressing issues. People were on the look out for any opportunity that earned them money. There were situations when people felt researchers could influence government to provide income generating opportunities. The other tendency was for people to take the problems they faced as individual households to be representative of what happened in the community at large. This is why (in the course of research, and serving as a session facilitator) I cross checked the gravity of the issue with other people. Participants in the meeting also provided rough estimates of people affected by a given problem in percentage form. This was mainly done during discussions with farmer groups, but was also attempted with other informants, such as district agricultural officers and young people (mainly 'bodaboda', i.e. those who transport people using bicycles and motor cycles).

4.4 The social system: community contexts for FFS

The previous sections focused on how technology interventions related to the existing farming situation. This section will give an account of who these farmers actually were and the extent to which their interactions around FFS can be explained in terms of social variables. The previous case by case structure is abandoned here, mainly to minimise repetitions, given that basic principles of social organization are similar across Eastern Uganda. Busia, Kumi, Tororo and Soroti districts are dominated by the same ethnic group (the Iteso). They are patrilineal. Women, land, cattle and children are considered the property of men. Men make most major decisions at both household and community levels. Male domination in decision making processes affects FFS meetings, and can be traced back to the cultural system prevailing in pre-colonial time. The kinship system is organised around clans, and males dominate all clan meetings and decision making (Emudong, 1974; Okalany, 1980). Women lack status or authority to intervene in their own right in the juridical domain (Webster *et al.*, 1992). Women are trained to be submissive and obedient to their husbands, and to respect male authority more generally (Atinyang, 1975). Inheritance follows a patrilineal rule, and women were inherited not unlike cattle. They could be used for example, as property to compensate crimes (Karp, 1978: 132). Out-marrying girls was looked on as source of wealth, since marriage required a bride price paid in cattle (Agemo, 1980). When school education began to spread priority was given to boys.

In the Teso community women and men play different practical roles. Some activities are meant to be performed by men only and some by women only. Men are supposed to provide all financial requirements for a home while women provide food and stay within the homestead. For this reason food crops remain female crops while grown for household use but become male crops when they assume a commercial value. In meetings, women talk less and sit separately, without mixing with the men. They relate more freely with other women. The community expects them to work with other women and not in mixed groups with men. Being male, age, leadership status, literacy and wealth are factors determining the respect a man enjoys in the community.

4.4.1 Social ties at group and community level

Understanding the operation of different social ties in a group and how they influence people's interactions provides an entry point for an understanding of prospects for participation in rural development. These ties include blood relationships (i.e. clan and family [patrilineage] membership), faith, management and use of shared resources (like water sources, fuel wood sources, swamps, pasture lands), and political affiliations. Burial of deceased family members, friends or community members is a strong social responsibility affecting all adult members of a village community. Death does not warn, neither does it visit a particular home. It is expensive, yet abrupt. Members of groups (whether based on kinship or association) tend to be called upon to give financial, moral and material support to the bereaved family/house. Women mainly contribute in kind (water, fuel wood, and food) while man mainly contribute to burial expenses in cash. Before FFS, farmers belonged to different self help groups formed around different connections and interests. About 41% of associations were formed around mobilization and use of local resources for mutual support Some then became FFS groups. The ties of socialization, labour, income generation and moral support basic to the emergence of these groups are explained below.

Socialization
In this context, farmers form informal associations for a range of reasons. For example, a group might come together to share tea in the evening or for prayer, but later take on other activities.

Prior to becoming an FFS under FAO-IPPM in 1999, Sihubira group began as a tea club with 4 men who met every Sunday in the evening. The tea group then started saving money to buy a bull for Christmas. This project attracted 10 additional male members. The 14 member group established collective commercial fields of sugarcane and a wood lot for (building) poles. With support from Christian Children's Fund (CCF), Food Security and Markets for Small Holders (FOSEM), and the On-farm Productivity Enhancement Program (OFPEP), members of the group embarked on crop production and became community trainers for improved production of cassava, beans, maize and soybeans.

A second example concerns Hulime and Abaringentie. Both started as church/prayer groups, and then later engaged in collective commercial production of cassava and groundnuts.

Pooling and sale of labour

Villagers, especially youth, sometimes come together to pool labour, e.g. in order to help members open up bigger areas of land, in time for the rains. Groups might then also sell labour to group members or even outsiders. Collective labour rotates around all group members in turn, and can be used for opening up land, planting, weeding and harvesting. Out of five groups of this kind on which I have data three (Buyaya, Busiime, Sihulawla) started as women's groups, one (Sikada) was formed by (male) elders and the fifth (Ndegero) began as a group of 7 relatives providing mutual aid. Four groups started when there was famine in the area in 1997. One group (Busiime) engaged in poultry production as a collective income generating activity; each member brought a hen and one member offered space to house the project. Sikada served as a rotational credit group, but the venture was tough for the ten old men who were members, given that they had no reliable source of income.

Income generation

Some members came together to pool resources (financial, labour and material) specifically to set up collective income generating activities. These activities included ground nuts and cassava cultivation (Mundaya group), raising turkeys (Asiaunut women's group) and drama/singing (Kyosimbaonanya). In addition, the drama group engaged in commercial vegetable production and livestock rearing (owning three cross-bred cattle under zero grazing for milk production and five goats for meat).

Mutual support (burial clubs)

Kawabona Kabosi FFS and Balonda groups first formed as mutual assistance associations for support in times of bereavement. But again, they took on other purposes than bereavement assistance, namely rotational farm labour, collective income generating activities (cassava, cotton, millet, beans and cotton) and saving to buy a bull for Christmas.

Loan schemes

Some associations were primarily rotational savings clubs (RoSCA). In these members contributed some agreed amount of money at regular intervals, either to fund loans to members, or buy utensils for members. For example one group (Buwumba) made contributions to offer loans both to members and non-members; interest for members was 5% and 20% for non members. Another RoSCA, Bakisa - later a FFS group also - started in 1998 with 10 members (5 husbands and their wives) contributing money so that members could in turn buy kitchen utensils (plates), pay their tax (men) and fund purchase of clothes for each family. They also pooled farm labour and contributed to buying a bull for Christmas. Lwala group formed specifically to be able to buy a bull at the end of the year.

Collective farming as a key social tie
The spirit of collective action can be traced back to the ties involving agricultural activity. Men and women once preferred working in separate groups, but the trend towards mixed groups is induced by requirements for gender balance in groups supported by projects. Some men actively seek to join women's groups, convinced that these have higher chances of obtaining financial and material support from government or projects, given donor concerns for "gender equality". Overall, the aspects of pooling labour and generating income from crops has crept through all the groups, although initial objectives steering group formation varied. Farm-related activities call for frequent and active interactions among group members. This tends to have the benefit of strengthening social networking and leads to more land being cultivated.

Drinking clubs
Sharing locally-brewed beer is a key social institution among the Teso (Karp, 1978). Alcohol connects people for purposes of leisure and entertainment, celebrating harvests, bride-wealth settlements, marriages and deaths, as well as offering payment in kind for labour services, throughout large parts of rural Africa (Bryceson, 2002). Preparation of beer is a women's activity. Although beer drinking was a commonly male practice in most parts of Uganda, in Teso region drinking local brew is cherished across age groups and by both genders. The brew (*ajono*) is made from millet. In the absence of sucking tubes inserted in a drinking pot and kept rotating from one person to another, a calabash is used. The idea is to minimize selfishness, strengthen social identity as I*tesots* (people from Teso) and build cohesion between friends and neighbours. The brew is present during all types of ceremonies, happy or sad, but is more readily available after the main harvest of millet and sorghum. Babies are initiated into the clan by inserting a drop or two into their mouths. Making of *ajono* has declined over the past few years as a result of reduced millet/sorghum harvests, mainly attributed to pests and weeds, especially Striga, and drought, as well as increasing competition for millet supplies for food. The cost no longer justifies producing lots of millet for brewing. Some farmers in Tororo claimed "it is cheaper to buy millet than to cultivate it..." Millet is now mainly produced for food. One further reason is that there is a wider variety of beer available on the market, including the bottled type.

4.4.2 Activities and roles by gender

Although activities and roles are often shared, in most cases women worked more than men. Studies show that women are over burdened in performing both productive and reproductive work (MFEPD, 2002; 2004), working for longer hours on a regular basis than men. Recent studies in Uganda have shown that men work slightly longer working hours than women on economic activities (Lawson, 2003), but where men's involvement in economic activities tends to be seasonal, women are involved in both domestic and economic, all the time. In Soroti (Abuket) and Busia (Lunyo), for example, both men and women performed the same activities, but with difference in intensity (Table 9). Men were asked to list activities in which they saw women engaged, while women were asked to list activities they saw men engaged in, in their home communities. Later the two groups came together and assigned an estimated percentage of time to each category of activity that had been identified. The percentages given refer to the proportion of women or men population in the village estimated to engage in the listed activity. For instance of all women in Abuket village (Soroti) those who took care of their homes were estimated to be 90%. That implied that for the 10% the responsibility of taking care of the home was either left to the children, husbands or to none. These estimations were collectively made by men and women in the villages of Abuket (Soroti) and Sihubira

Table 9: Distribution of activities by gender and percentage.

activities in which the community is involved	Soroti (%)		Busia (%)	
	Men	Women	Men	Women
Land preparation				
• ploughing with oxen	80			
• slashing (mainly for cash)			90	
Buying sauce (meat and fish)	20			
Caring for the home*		90		90
Meeting medical and education bills	90			
Looking after cows	80			
Ensuring security at home	85			
Farming for income generation	95	20	70	
Farming for home consumption	30	100	40	90
Looking after small ruminants and poultry		40		70
Post-harvest handling		90		90
'Complaining, quarrels & rumours'		90-95		50
petty business or trade (fishing and fish mongering, selling local brew, produce, markets, shops, hotels, local bread, boda-boda work)	30	20	60	40
Drinking local brew	50	30		
Participating in groups & community activities		40		80
Civil service (e.g. teaching)				20
Smuggling items across kenya-uganda border				35
Others (hair dressing, playing cards, building, carpentry)			20	08

*caring for the home includes looking after children, preparing food, caring for the sick, ensuring availability of water and firewood and general hygiene as well as purchase of household equipment like plates, cups, cooking pots, furniture, clothing especially for the children.

(Busia). Each gender was looked at differently, not in comparison to the other. That explains why the summation of proportion of the men and women engaged in the same activity do not add up to 100%.

Land preparation and income generation to meet financial requirements of the home were mainly activities performed by men. Income generation was mainly through cash crops like cotton, groundnuts, millet and maize. Women were mainly engaged in ensuring food availability and comfort for the home. Post harvest handling of all crops and participation in group and community activities especially at funerals, marriages and other ceremonies, were mainly women's activities, although some men took part in these activities too. To ensure food security in the home, women had a variety of fields under different food crops like sweet potatoes, maize, millet, sorghum, cassava, groundnuts, cowpeas. Apart from cotton, men had fewer but bigger commercial fields under the same food crops. Although the men rarely lent a hand in women's fields, women worked in the men's fields. Many times, men's fields were given first priority, since these were the major sources of income for the home.

Although men mentioned "complaining or quarrelling" as one of the major occupations of women, it was then observed that women often engaged in such activity as a way of expressing their disgruntlement with household activities and decisions taken by husbands resulting in absence of money, relish, and clothing, or development of extra marital relationships with other women, where resources drained away to the disadvantage of the wife. This was perhaps one of the few weapons at the disposal of women, given that they have very little influence over decisions with regard to use of money, despite contributing greatly to its acquisition.

Why more women than men in FFS groups?
Both wives and daughters spend much time in the commercial fields of husbands and fathers, in addition to being busy in their own food fields. The women do much harvesting, drying, storing, threshing and packing prior to selling but rarely benefit from their sweat financially. Because of this, they feel exploited by their husbands and deeply locked in poverty. This could also explain why FFS groups had more women than men. Women are currently on the look-out for any opportunity that builds on their knowledge and skill to be able to engage in feasible income generating activities. The more a crop dealt with by FFS was perceived as commercially oriented, the more the number of men involved.

Vulnerability and lack of resources make women proactive in initiating self help groups to provide mutual comfort, advice and support. In spite of providing around 90% of agricultural labour, women lack control over household income, even when it results from crops they have produced themselves (UPPAP, 2002). Figure 6 offers some specific data on gender imbalances in FFS groups. Factors explaining this imbalance include: (a) a majority of old groups started as women's groups, (b) women are economically and socially disempowered, as observed by PEAP (2004: 19) and search income generating opportunities/activities via groups, (c) women enjoy learning together without fear, especially when they perceive themselves as belonging to the same class of poor and marginalized people, (d) The responsibility to feed family is left entirely to women, therefore they thirst for knowledge and skill related to increased food production.

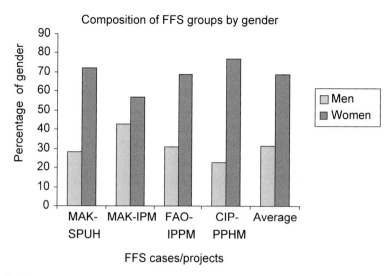

Figure 6: Gender composition of FFS groups.

Informal enquiries from men and women revealed some further reasons for lower male engagement in groups. Many men perceived groups as "collections of poor people". Men also feared embarrassment if they showed they did not know something, especially in a group of women (their attitude was 'I remain the boss and know it all'). Some were away from home for long periods looking for money. There were more chances of making money in urban centres than staying at home and engaging in agriculture as a commercial activity. Indeed as confirmed in a recent study on the impact of PMA/NAADS (Oxfam and FOWODE, 2004: 2), men simply cannot afford to spend hours on training with no immediate financial gain.

Men tended to engage more in out-of-home activities. The major activities that men in Busia engaged in to generate income kept changing depending on the situation. For instance, cotton was the main cash crop but when the market dipped most of them resorted to trading in coffee, buying it from neighbouring districts in Uganda (mainly Busoga region) and crossing to Kenya. When coffee prices fell, many men then resorted to fishing. The Government of Uganda has established strict rules to minimize indiscriminate fishing, which has since reduced the number of people in this business. Trading in smuggled goods like oil, fuel, cigarettes, cloth and mattresses has boomed, even involving school children at some point! Stricter rules and the death of some people at the hands of the Uganda Revenue Authority (URA) has made the business more risky, and people now shy away from it in love of their dear lives. However, many acquired capital from smuggling and opened up retail shops in active trading centres. The other activity that many men now engage in is farming for cash, specifically maize, groundnut (Serenut II) and bean. Rice was mentioned to be a promising crop. Millet, cowpeas and sesame had attractive markets, especially across the Uganda-Kenya boarder, but were very labour demanding and time consuming. Hired labour was scarce and expensive - reducing profits.

4.4.3 Influence of social variables on the dynamics of FFS

An understanding of how men and women relate guides project implementers in thinking through the projects, effectiveness in set up, who to address and how to address them. Interaction between men and women in FFS has implications on whether FFS targets the most appropriate people. Although women outweighed men in number, men dominated most discussions and activities in FFS. On arrival at FFS sites, it was clear to the researcher how even seating pattern and working in mixed groups had effects on gender relations as will be seen below.

Arrival at FFS site
Because most farmers gave priority to their individual/home gardens, FFS sessions were shifted from morning to afternoons to fit farmers' schedules. During FFS training men came in earlier while more women came in later. The males had less work to do at home. Most of it was done by women. From the garden, women had extra work to prepare homes (e.g. fetching water, cleaning, cooking for the people left at home) prior to joining FFS. Cumulative tiredness as a result of unbroken work without adequate rest probably explains why women often dozed through the sessions. Male counterparts did not appreciate the reasons behind women dozing off, but instead made fun of them, "It is old people who tend to slumber..." But perhaps they were also bored, feeling disengaged from discussions, since women rarely talked.

Many times, the chair and executive of the group arrived earlier on site. This enabled them to tell who was absent and reach him or her in time. Once the facilitator had arrived, the chair person went round with a bicycle reminding the rest to be present. In this kind of mobilization, preference was given to a specific category of people: literate, talkative and therefore perceived to be very active in FFS discussion. These were mainly men. Cultural traditions also partly contributed to the bias against women actively engaging in discussions. Traditionally, it is not right for a married woman to be picked by another man. This explains why men were more comfortable looking for and calling other men.

Seating arrangement
In encouraging men and women to sit together, FFS seemed to be violating a social norm! A discussion with some elders (both men and women) in Soroti district revealed that women were not supposed to sit in the same group with men. Men are supposed to sit on 3-legged stools and women on the ground in a separate group. While in the FFS class, men sat on chairs on one side, mainly at the back, and women sat down on mats or pieces of cloth on another side, or at the front. This fits the culture that gives respect to men (husbands and fathers) and ensures in-laws of the opposite sex are not inconvenienced. In Uganda, especially in the central and eastern region, it is a taboo for wife/husband to sit close to his/her father/mother in-law. I experienced this when I requested members to come closer and one lady preferred to keep a long distance from the group. Little did I know about the relations of group members! "I am an in-law..." she explained, as she pulled her mat further away from where the father in-law was seated. Some female local leaders sat on chairs. The seating pattern (chairs and mats) thus

made a point about power relations as well as gender inequalities - those who sit on chairs are more powerful and influential. Across both gender, the literate and economically more prosperous group members sat close to each other and talked more frequently than the rest during group discussions.

In rural communities, better clothes are preserved for church or for travelling to town. Wealthier farmers either had better clothing or could at least put on sandals, while the rest of the group went bare foot. These may not be accurate parameters to identify the 'least poor', because some people just did not mind how they dress or appear in the eyes of others, but it is a rough guide nevertheless. In some FFS, especially where there were more young ladies, poorer members tended to view the rich as proud. This was psychological, but had the effect of sapping morale to participate. "Those who are richer think they know more or better than we do and are very proud; they talk a lot in order to be noticed..." said one lady who kept quiet all session long.

Contributions during group discussions
During FFS sessions, men talked more than women, yet there were more women than men in the groups. Being better informed or more widely travelled perhaps explains the activity and confidence among men. Cultural stereotypes portray men as bold, and taking the lead in discussions fitted this image. It was therefore up to the facilitator to be alert to the need to bring others, especially women on board, and reduce the chances of quiet and marginalized people giving up. Calling farmers in FFS by name was one way of valuing them as people contributing to group learning. Reference to the executive members by function (as chairman, vice etc) and the rest of the members by name affected the freedom of the rest of the members to make contributions. This suggested that the executive was on a higher level than the rest of the audience. Fear to challenge the executive was a likely outcome, which might be constructive or destructive of group performance, depending on the motivation and capacity of the executive. But in general FFS meets its objectives better where it engages the interest and contributions of all. Encouraging all participants to respect and take each other as equal (no titles, but names for all) increased and eased interaction among members.

In one FFS under the A2N-ISPI project in Tororo, and in another under CIP in Soroti, ideas from old women were often laughed at with excuse that they were out dated! Such treatments, especially from young men, embarrassed the ladies. It minimized active involvement of this group in class discussions, and the FFS thus lost the chance to learn from their experience. This loss was not negligible when it is recalled that women are more involved in agricultural activities than men, and that old ladies have an especial wealth of experience. But in any case laughing at old people contradicts an Iteso cultural norm, in which respect is directly proportional to age. In this case, older men commanded greater respect, showing that respect or superiority is dependent more on gender than age. Confidence and freedom of expression developed by some FFS female members was often mistaken by men as "being argumentative", therefore behaving contrary to the cultural stereotype that a woman is supposed "to be quiet and always submissive".

Gender mixing

FFS facilitators emphasized a gender and literacy level mix in sub-grouping processes for Agro Ecosystem Analysis (AESA) exercises. AESA is the principle tool in FFS training where farmers undertake a series of activities with guidance of a facilitator. One, make field observations to monitor crop performance. Two discuss and present their observations (in pictorial form on a flip chart) and suggested actions to be taken. Three, present the outcomes to the entire group in plenary. And four, engage in a general discussion from the entire group about what to do next based on available information (observation and experience).

Some facilitators decided who should be in which sub-group to ensure the balance and complementary of memberships. This was contrary to the old social tradition in which women are not supposed to mix with men or speak but listen and comply with decisions made by men (Administrator Teso Cultural Union, 2005 *pers. comm.*). Division of the big group into sub-groups increased opportunities for every member to participate in the FFS training activities. Even the quietest people spoke a bit during the sub-group meetings. There was more cohesion and friendliness among members in the same sub-group as compared to the interaction across the entire FFS group. As a result of the closeness, some members of sub-groups felt it an obligation to attend FFS training not to disappoint their sub-group members. "My group members feel let down when one of the sub-group members fails to turn up..."

In situations where some members were unable to turn up due to sickness, funerals, or attending to other priorities, they sent non-FFS member to represent them. Often, mothers delegated their children to attend on their behalf, with the assumption that the children would teach them what they missed. Children are often excited at narrating what they observe. One lady in Busia delegated her niece to attend FFS training, but the facilitator did not welcome that. "Tell her to come herself but not to send anyone else..." said the facilitator, with eyes fixed on to the young girl, who bowed her head down in shyness. Interacting with children in the same group would mean treating them on the same level as adults, but traditionally children are never taken to be people with constructive ideas, especially among adults. They are subordinates, with no autonomy of thought or action in adult matters. Although delegation implied commitment to the task, it jeopardized consistence in attending FFS sessions, hence interaction/learning about the technologies. To the children, the arrangement was completely new, and they could not easily fit in. The chances of elders taking up their contributions seriously seemed very minimal. Nor could child delegates understand linkages between previous and current sessions and thus tended to keep quiet all through. On the other hand, involving the younger generation would mean application of most things learnt, since they are more anxious to experiment, and are many times working with their female care-givers in the fields. Thinking about ways to incorporate children positively into FFS sessions is a worthwhile challenge.

Men preferred belonging to separate sub-groups from women, reasoning that "women are slow and not as bright..." (a thought mentioned by some men in A2N-ISPI- FFS groups). This shows that men were not comfortable discussing and presenting in the same groups with women. Discomfort among men could be attributed to their limited experience and therefore knowledge in crop related issues, on one hand, since they spent less time doing field work. On the other hand, the tendency of seeing their status seemingly undermined especially when challenged by women is another likely explanation of the discomfort. If it is in fact true that women are slow learners it could be that they were more analytical than men, and took their time to understand in order to make a better decision. But this certainly meant that being with women in the same group involved staying longer in FFS sessions. Men were often in a rush to finish up things. One male dominated sub-group in one FFS spent less time in the field during AESA observations. Some members even began writing/drawing what to present before actually making observations! Challenged, one said, "These are the things we do everyday..." To them it was more a routine than a learning opportunity: after all it is a woman's responsibility to ensure availability of food in the home. Most of the time, men discussed how to make money and politics than actually attending to the requirements of AESA. The tendency of this sub-group being "very fast" then created pressure on other (more committed) sub-groups, who felt they were too slow. The issue was then not the AESA itself but how fast to do it through visuals. This sends a message that most of what was presented during AESA may not have been observation-based learning at all, but the enactment of an assumption by a group of under-motivated men about what they expected to happen. This will not be the first time that "the boys" in a science class have failed to notice what was before their very eyes.

Men were not very comfortable mixing or talking with women (especially their wives!). Companionate marriage is still quite rare in the Ugandan village. Talking to or with a woman was reportedly regarded as an inferior action in Busia. "When people see a man talking with his wife in search of an agreement upon something, they call him stupid and undermine him..." three women mentioned in chorus during a group meeting I held with them. The man is expected to be decisive and tell his wife what appertains. A man who discusses with his wife is little respected by his fellow men. Only the God-fearing men worked with and talked with their wives. The other men referred to them as 'chicken' - i.e. ones who return home by dark, along with the chickens. The implication was that real men hang out drinking with colleagues and just instruct their women folk, rather than seeking consensus with them. The number of women gaining confidence to influence decision making processes at household and other levels, however, is slowly increasing due to greater exposure opportunities. Such women visibly contributed constructive ideas during meetings, suggesting that FFS in Uganda will "mature" towards fuller potential with time and usage.

Gender and power relations among FFS farmers
Leadership in a group or community was known to belong to the most powerful in terms of knowledge, exposure, position, wealth and education. This explained why men took up most leadership positions. Women mainly took up the position of group secretary - confirming the mentality that secretaries are meant to be women. In case of an invitation for a meeting or training that needed one or two group members to represent the group, the men always took the opportunity. Absence of women (mothers and wives) from home for long hours implied confusion; children might go without food and a bath, and the house and compound might remain dirty in the absence of women at home. The notion that men and women might share household chores has a long way to travel in rural Uganda. But on the other hand women at home were always available and probably a reason as to why their attendance in field schools was

more regular than that of men. Some men found it unavoidable to join women groups since the women kept the meetings rolling along while men went to town to hunt for money and jobs.

In purely female groups, men still influenced the executive, especially the chair, in making decisions for the group. Husbands who had financial hopes invested in particular women's groups advised their wives (as chairpersons) on what to do in the group so as to win group support. Most decisions from such female leaders were mainly made under the influence of husbands. In one group under FAO-IPPM in Busia, one husband convinced his wife (as chair of a women's group) to sell off the group assets for the benefit of the family. Women who can express themselves, especially those able to speak some English, tended to lead the rest of the members. In most FFS related meetings called at the district level, however, more than 70% were men. In such meetings, women rarely spoke. They always kept quiet and waited for the men to make contributions.

Gender inequality is a social phenomenon emerging from set task divisions in economic activities. It has a general dimension, but also culturally specific aspects. Working on these specific factors will have a positive impact on activities influenced by gender inequality. Agriculture is one major activity whose development is influenced by gender relations. Women greatly support the sector and any support to them will boost agricultural and rural development. Given that there were more women in FFS activities, it would be good to develop specific strategies for turning FFS opportunities into tools to empower women. The empowerment process requires access to and control of economic and social resources, including supportive policies in regard to land and credit, among others. Rooted patriliarchal relations and patrilineal inheritance norms remain among the biggest obstacles to women gaining land ownership rights (Kasente et al., 2002). Lack of land rights affects whether or not technologies can be implemented. FFS needs to develop a perspective on agrarian institutions, such as land tenure and land renting contracts accessible to women, if it is to empower women through innovation policies (cf. Walaga et al., 2000: 36-37). Strong political will and supportive agrarian policies are needed, besides adjustment in the legal and cultural status of women, given that many agrarian activities in Uganda are largely feminine (Oxfam and FOWODE, 2004). FFS and other related participatory approaches can play a role in lobbying for action on e.g. the domestic relations bill in Uganda[22] as a key aspect of technological transformation in agriculture, just as training women in entrepreneurial skills will enable them engage in independent commercial activities through which they will gain the resources to purchase land and other assets.

4.5 Concluding remarks

Farmers do not take up technology for its own sake. Technologies developed and chosen by researchers do not necessarily fit farmers' local realities. The improved crop varieties that are

---

[22] The domestic relations bill is an amalgamation of all domestic related laws that would give women equal rights to men in making decisions in homes as well as in sharing property including land. These laws include marriage, divorce, separation, and inheritance and property rights. They have been ignored for the last 30 years

the major technological component of FFS do not contain features that were most valued by subsistence farmers. To most farmers, taste (edibility, palatability, sweetness) takes priority over other characteristics such as yield, size and resistance to pests or drought tolerance. FFS agronomic practices deviated too far from what most farmers were used to, and needed more investments (of labour and time) than farmers could afford. Most interventions were better suited to commercially oriented farmers whose priority was income generation as opposed to the subsistence farmers whose priority was food. This kind of FFS package as was introduced in Uganda appeared more suitable for commercially oriented farmers (mainly men) than subsistence oriented farmers (mainly women), and yet women dominate the composition of FFS memberships. Women see FFS membership as a way of developing a cash income stream of their own, but lack resources and capital required to put into effective use the knowledge and skills they acquire from FFS.

Learning what works and how to improve on its effectiveness in a given farming system needs to be taken more seriously if technologies are to make a difference in the lives of Ugandan farmers. Understanding local practices provides useful pointers towards choice of effective interventions, feasible and relevant to specific community contexts. Local adaptation is crucial, because farmers find it easier to take on or engage with interventions that build on or are compatible with existing knowledge, skills, labour and production goals. Understanding farming practices and the rationale behind them is a challenging and gradual process, however, requiring patience, commitment, and time. Open mindedness and the desire of researchers to learn from farmers is essential.

Gender is not a neutral factor, and how men and women relate affects efforts to link farmers' capacity with introduced technology. A patriliarchal culture in which men control all resources and take the lead in making decisions remains dominant in Uganda. FFS does not, in practice, manage to overcome this cultural mind set. How might we more genuinely support marginalized women? For FFS to link effectively with farmers, it is important to identify priorities affecting different categories of farmers, men and women, young and old, richer and poorer, and seek to build the capacities of all relevant groups. It is unclear that an approach based on existing community values will be able to fulfil such a task. FFS may need to develop a class analysis, and target social obstacles to the empowerment of the poor, if technology transformation is ever to succeed.

This chapter has shown that farmers, men and women, are not passive receivers of technologies but eager experimenters. As argued in the introduction, to understand the learning component of FFS requires making a distinction between learning and teaching. As Lave (1995) observed, teaching becomes successful when an effective link between different contexts is made. Such connection can only be effective when learning takes place among all members. Teachers thus become learners and vice versa. From that perspective the FFS projects discussed in this chapter certainly were not a great success. The gender elements discussed here, however, show that FFS does have an impact on social relationships. From the theory the move forward would be that an FFS has to be adjusted in such a way that learning effects increase among farmers and facilitators, men and women.

CHAPTER FIVE

Local organisation of FFS: from curriculum design to functionality within wider structures

5.1 Introduction

In the previous chapter we discussed who the communities of learners were and the extent to which technologies promoted in FFS matched the existing farming practices and social contexts in project areas. In that chapter some elements of an FFS, like the Agro-ecosystem Analysis (AESA) and group meetings, have already been mentioned. In the present chapter the focus is more on the FFS package as whole. We will look at 'FFS-in-action', including organisational features on the ground and how that relates to other local initiatives and structures higher up in the FFS organisation. We will also consider the connection between FFS and "neighbouring" grass-root activities. A central feature of an FFS is the curriculum. The content, design and execution of the FFS curriculum reveal something about how relationships between features on the ground and those at higher levels are established. This chapter gives an analytic description of features known as agro-ecosystem analysis, energizers, field tours/visits and field days, as the main tools through which the lower and higher level structures are linked. It also describes the linkage or integration of FFS with other local structures and activities.

The reason to look at these elements is to consider how learning effects, as intended by the introduction of FFS, are enhanced (or impeded) by the curriculum and how other factors, not taken up in the curriculum design, affect the learning process.

5.2 FFS curriculum and the Agro ecosystem Analysis

The FFS curriculum as applied in the projects analysed in this thesis was developed by scientists in research and academic institutions. Scientists adapted a standard FFS curriculum format to suit the technologies to be transferred in the various projects. Under MAK-IPM and MAK-SPUH for instance, feasibility studies were set up to get insights into the farming situation of targeted project areas. This was basically consultation with district officials and other key stakeholders (e.g. FAO staff), which was then followed with piloting and full implementation. Prior to establishment of pilot phase, pre-test studies in form of surveys (questionnaires) were conducted to ascertain farmers' knowledge on specific content of interest to the project implementers (e.g. IPM in groundnuts and cowpeas for the MAK-IPM case, pesticide handling for MAK-SPUH, and soil management for A2N-ISPI). This has been interpreted in chapter two as evidence that research mandates from higher levels shape the operation and outcomes of FFS. The IPM principles of growing a healthy crop, conserving natural enemies and conducting regular field observations upon which IPM-FFS is based, point towards reducing use of pesticides while at the same time managing pests in the field. An ideal FFS curriculum (Box 9) is emphatic about the technical skills needed in managing

Box 9: Standard IPM- FFS session schedule.

FFS sessions usually include the following activities
1. Opening with greetings and prayers
2. Recapitulation of material covered during the previous session
3. Introduction of the day's program
4. Agro-ecosystem Analysis (AESA): In groups of five, with some observing IPM plots and others observing non-IPM plots, participants observe general field conditions, sample ten plants, collect insects, make notes and gather live specimens. Each group makes analysis of its observations and analysis via visual means (AESA drawing). AESA drawing includes pests, natural enemies, weather conditions, plant condition, field condition and action decisions. A member of each group then presents findings and rationale of the group analysis to the larger FFS group during an open discussion. The discussions lead to a consensus decision about what to do next. Decision points include need or no need to spray, continue with field observations, setting up some experiments on pest-predator dynamics etc. Participants need to understand the process and purpose of AESA
5. Introduction of special/new topics. This is linked to stage of growth of the crop and specific local issues. Special topics may include pest control, crop physiology, health, field ecology, economic analysis, water management, fertilizer use, etc.
6. Group dynamics/energizers: various activities are undertaken in relation to the special topic of the day or in regard to any local (farming) problem. The process is synthesized so that participants identify key learning points and learn from the experience
7. Evaluation of the day and planning for next session

Note that for every activity, the role of the facilitator is to help participants learn or discover but not to teach.

pests and looks at agro-ecosystem analysis (AESA) as a cornerstone in building farmers skills (Pontius et al., 2002: 21).

The centrality of AESA lies in its linkage with improved farming systems. Farming systems cover the full spectrum of social and technical aspects influencing farming practice. Farming Systems Research (FSR) is generally considered a common research approach, and includes a set of standardized methods to analyse agriculture. A curriculum is a set of courses with focus on content. An FFS curriculum is a set of a series of sessions following the phenology of the crop in question. The number of sessions therefore depends on the maturity period of the crop. For instance, a crop like groundnuts or cowpea had about twenty sessions while sweet potatoes had about forty sessions. These sessions began from site selection to post harvest handling. Central in the sessions was AESA.

When performed well the AESA (in the IPM variant) results in a plan-for-action that has the intended outcome of reducing losses due to pests. Experimental plots (five by five metres and five by ten metres, depending on availability of land) are established with different treatments (technology packages) providing a central focus for field observations. Farmers make plot-based field observations for data collection, presentations and discussions of performance of technological treatments. Depending on the topic, treatments in the experiments varied from single to combined application of a variety, planting methods, spray regimes and types of fertilizers. In sub-groups allocated to different plots with varying treatments, farmers collected data during observations of the learning plots. AESA worked out differently in the different projects.

In one FFS dealing with groundnuts under MAK-IPM[23] farmers were randomly divided into three subgroups, each making observations and collecting data on one type of treatment for all the five groundnut varieties, i.e. three improved varieties of Serenut I and II and Igola I and two local varieties (one called Red Beauty and another called *Kabonge*). The three treatments per variety were no spray (the control), spray twice (recommended IPM practice) and spray four times (standard practice). Farmers had no sub-group analysis, presentation or discussions. Each farmer in a plot randomly chose 2-3 hills to observe for insects and reported observation directly to one person (often the facilitator) who took records. A record of insects was based on the average of the samples made per plot. The objective of taking the average number of insects was mainly to bring out the dynamics in pest population as the season advanced. Some insects and plant disease (pest) symptoms unfamiliar to farmers were meant to be taken as live samples to scientists (especially entomologists) for identification with the help of facilitators. This however, hardly happened because of costs and inconveniences to the facilitator. What was not known to the facilitator and farmers remained unknown to all. Very little attention was paid to the agro-ecosystem as a whole. The sessions were based more on teaching than self-discovery, and take-home messages tended to emphasise the superior resistance of improved varieties to pests and drought, when compared to local varieties (the comparison were often based on evidence from on-station research), and the benefit (in terms of cost reductions) realised through spraying twice instead of four times. The comparisons seemed to be based on evidence from research stations

Under A2N-ISPI, sub-groups of farmers in the FFS observed different plots. Depending on the number of crops and treatments, sub-groups observed, discussed, made presentations, both in drawing and verbally, for at least one plot (if there was a single crop under study). In Karwok FFS (Molo sub-county, Tororo district) four sub-groups made AESA of at least three plots per study crop (maize and groundnuts). Improved varieties were sown on each plot. Maize treatments included applying farm yard manure (FYM), inorganic fertilizers, against a control with no soil fertility treatment. Treatments to groundnuts included FYM, compost, single super phosphate and a control. In some cases, the same sub-group observed the same plots throughout and in other cases, the sub-groups rotated on a weekly basis between different plots. This gave equal chance for everyone to take an active part in carrying out field observations of all plots. All sub-group members were expected actively to carry out

---

[23] Under the MAK-IPM FFS project, there were two FFS groups in Iganga district, one in Bulamagi sub-county and another in Nakigo sub-county. I was a facilitator of one FFS (Buwolomera, in Bulamagi sub-county., the basis for a number of the direct (as opposed to reported) observations in this thesis.

observations, then discuss and make an agro-ecosystem picture analysing what they saw on a flip chart ready to present findings to the other farmers in the entire group. This did not happen in all groups. In some cases a few people took the lead, and did the task alone, but there is reason to believe this was on behalf of the other members of the sub-group. Observations, therefore, were not confirmed by other group members. In most of these cases these leaders were either men or more informed people, especially in mixed groups where women often held back. Field observations were based on ten plants randomly selected per plot along the two diagonals for a general representation of the condition in the plot. Some members did not even bother to carry out field observations but made the drawings first, and then made field observation as a routine gesture to the rules. Their reasoning was that they knew what to expect in the field, and that observations were a waste of time. In this case, AESA seemed to be reduced to drawing.

Under the CIP-IPPHM, MAK-SPUH and FAO-IPPM projects, farmers also carried out AESA in subgroups. Observations were based on ten randomly selected plants along two diagonals of a plot. Following the diagonals was observed to be more common in extension staff-led than farmer-led FFS. The random collection of samples was a requirement for a statistical analysis by researchers. Across all projects, data collected was mainly quantitative. It included plant height, number of leaves and leaf/shoot lengths, insects and animals present (types and numbers), number of roots, and yield. AESA intended originally as a participatory observation and analysis exercise was thus turned rapidly into a collection exercise to generate quantitative data. Failure to highlight interactions leading to quantitatively expressed observations limits the potential of an AESA to support improved understanding. This may explain why many times farmers collected information but seemed to have little idea about the importance of the underlying processes or even what purpose would be served by the information collected.

Some farmers not only had little idea what to do but no real understanding of why the various measurements were made, as confirmed (in chorus) by four women farmers in one Asianut CIP-IPPHM FFS group in Soroti: "We do some things without knowing why and little is explained..." In this case, attending FFS training became a routine or compulsive activity instead of a process which resulted in increased understanding. In these conditions, farmers sometimes only walked along the edges of the plots, and had little motivation to bother themselves by moving inside. The overall impression gathered by the researcher was of farmers failing to understand the organization of the experiments, and the importance of field observations and data collection, as applicable to their own farming contexts.

Across all projects where AESA was conducted in sub-groups, save for MAK-IPM, each sub-group recorded its observations on a flip chart in pictorial form and one member presented to the rest of the larger group the findings of her sub-group. A picture of the crop being studied was drawn in the centre of the paper (using green crayon) with insects on the sides. One side had pest predators and the other had insect pests (in different colours). Pictorial form simplified understanding by illiterates but needed some skill to be able to draw the exact appearance of a given organism. Farmers did not have local names for some insects and used the English or common name, as learnt from the facilitator. Day of planting, number of the different types of insects and date on which observations were made also appeared on the chart. Each group member was given a chance to present. The rest of the big group asked questions. Sometimes these were genuinely for clarification, but in other cases seemed more an opportunity to display status (i.e. authority to speak in public gatherings, often a marker of male prestige). Collective decisions about what to do were addressed after all sub-groups had presented. In most cases, there were situations where no collective decision could be reached, but few farmers looked uncomfortable. All they knew about the purpose of the operation was making field observations and presenting them. Some of them continued discussions on their way home, but did not bring their concerns back to subsequent plenary discussions. The flip charts were subsequently folded up and kept at the chairperson's place. They were shown to visitors who requested to see them. These flip charts were a rich potential resource that could be used as reference for farmers to link the previous and current field observations, to get better insights about the need for regular field observations with reference to dynamics in the agro-ecosystem. This would update members who were absent the previous week and make farmers see the patterns of interactions and outcomes of processes across the sessions or weeks from planting to harvesting. The fact that no such usage was encouraged suggests that it was not only farmers who were unclear about the overall spirit and purpose of FFS.

These observations about the way in which AESA was conducted suggest a deviation from the original FFS aim of building farmers' observation, analytical and decision-making skills. Neither facilitators nor the scientists who trained them seemed to have fully internalized the processes and objectives of AESA and therefore could not provide adequate guidance about the process to farmers. AESA activity was tailored more towards data required by scientists than towards helping farmers understand the important field processes AESA was intended to reveal. These observations offer a picture of FFS methods turned into standard science-led field research practice. In sum, adopting and reproducing the IPM-FFS curriculum as a finished product undermines the FFS objective of building up farmers' capacity to manage a particular problem through locally-acquired hands-on learning.

5.3 The internal organisation of FFS

This section looks at how FFS was organised internally and how the interactions within influenced operation of FFS. How AESA was handled, use of energizers and how funds to run FFS activities were management, are the three main activities described and analysed below in understanding the internal organisation or interactions of actors in FFS. A general perception of FFS model is also given as we will see.

5.3.1 How AESA was handled to build farmers' analytical skills

Process and outcome are equally important in managing crop production or farming practices through AESA. Focusing on the process is more educative than just observing the results, and contributes more to the building of analytical skills. The idea of AESA is to empower farmers with knowledge and skills to make informed and appropriate decisions in their own farm management. In this section some further details are given about how analytical and decision making skills were supposed to emerge, and what the outcome was in reality.

Design and establishment of experiments was often done by facilitators and in some cases by a handful of farmers, especially members of the executive. This left out the majority of farmers from the process right from the beginning. It explained why some farmers could not state what treatment had been applied to specific plots, and what the treatment was supposed to show. For instance under A2N-ISPI area of experimental plots and experiments were established by the facilitator in one FFS (in Busia) and a handful of farmers took part in the establishment of experimental plots and treatments in another FFS (in Tororo). The appearance of the experimental plot determined the interest with which farmers carried out field observations and data collection. In most situations, farmers preferred plots with healthy looking plants promising superior performance. They rarely liked to look at, and actually dodged making observations in, miserable looking plots. Lack of guidance in analysing these "miserable" plots denied farmers a chance to understand the reasons behind (extreme) variation in plot vigour in terms of agro-ecosystem analysis. They often based their reasoning about "bad plots" on instant judgements about climate or the poor fertility of the plot.

There was very little discussion after the sub-group presentations to help improve upon understanding about interactions of elements within the agro-ecosystem leading to observed outcomes. Various claims were stated about assumed reasons for crop performance but this hardly led to a ranking of the less and more likely explanations. According to observations on A2N-ISPI and CIP-IPPHM, it transpired that status differences determined the outcome of the discussion. Prestigious and verbally strong farmers overshadowed the contributions of the rest, especially the illiterates. It was also an opportunity for the less talkative to keep within their comfort zones (remaining quiet and observing others). Translating this in terms of organizational principles tells us something about the way institutions think and the need to understand institutions (Ostrom, 2005). Organizational processes within a social group reflect the thinking of its members and vice versa. Likewise, Forgas and Williams (2001) remind us that influence is more frequently indirect than direct and is often not directly observed or noticed. Discussions focused on pests and performance of introduced varieties in the experimental plots. Personal experiences in relation to content of discussion were rarely explored across all projects. Argumentative discussions, coupled with the fear of others looking at some practices as outdated, as was observed in one A2N-ISPI FFS session, contributed to unwillingness to share experience. The facilitator too often either forgot or ignored citation of experiences from members present.

The debate format made some members, especially women, shy away from the presentations, leaving the literates (mainly men) to make the sub-group presentations most of the time. This minimized the wealth of women's experience, an important loss given that women were more involved in farming and generated more experience than men. A debate that drives through discussion and ends with dialogue enhances interactions more effectively than a discussion that is skewed towards debating. Facilitators did not adequately guide the discussions for fruitful decisions mainly because they did not know how to do it. The only way they did it was the way they had been taught during their training. They were at a loss to devise appropriate ways to engage the different categories of people, and overcome imbalances of discursive capacity reflecting differences of gender, social status and wealth.

Farmers' perception of AESA

Based on the way AESA was conduced and how they were involved, farmers developed different perceptions about the method. To most farmers, AESA was understood in relation only to insects. Their facilitators emphasized and limited the discussions to insects, as it was emphasized during (IPM-oriented) training. Farmers appreciated AESA because it increased their knowledge base, especially with regard to insects. The very active and vigilant farmers learnt to identify and differentiate pest types, pest dynamics, pest effects, difference between harmful and beneficial pests. They learnt that different pests attacked the crop at different times. Some insects increased in population numbers and new ones came in as the crop progressed. " whiteflies and jasids, leaf roller and miner beetles, come in earlier while ladybird beetles, millipedes and rats come in later..." Farmers however found AESA very bothersome especially when it came to "academic" activities like counting insects (pests and natural enemies) and measuring the lengths of cobs and panicles (see below). To them, estimates were handier. Although counting and measuring gave a more exact estimate of a parameter or variable, the exercise did not seem farmer friendly. "Do farmers really need to count pests and measure heights to understand interactions in agro ecosystem?" was an implied question.

The academic nature of AESA. According to farmers in FFS, AESA needed some standard of education given the debates and arithmetic involved. Most participants were illiterate and often felt inferior or foolish. They did not like appearing in such a bad light, and opted to shy away or escape when it came to making presentations. Drawing proved difficult, especially for the women and old people, who had to be taught how to handle a pen. To avoid embarrassment, farmers in this category opted not to take part in drawing or writing results on a flip chart. But it is important to note that they were very active in fieldwork and in discussions held at a more informal level. Arguments, blaming and debates during presentations and discussions often caused tension between members. "I fear to present the AESA in plenary because people will laugh at me and ask me hard questions..." said one female caught going away prior to presentations. Literates - mainly men - tended to pin down illiterates, who were mainly women. This contradicted with the principle of adult learning that focuses on mutual respect for interactions. Strong and positive self esteem is conducive of healthy growth, development and interactions and in skill building, but the disdain of the more educated tended to undermine the process.

Dominators or the "big talkers", as Chambers (2002: 180) refers to them, limited the opportunity to hear and learn from others' experience or opinions. The silent ones lost the opportunity to talk. Some authors see a solution in creating a relationship of mutual respect and support for optimum sharing of knowledge and skills (Whitaker, 1995). Richards (2006) however shows that the emphasis on discursive skills in participatory approaches creates a blind spot for non-discursive performative expressions of farmers. Farmers often best make their points through activity in the field. Those who do not have the skills or courage to stand up and speak in FFS sessions thus might make unnoticed "statements" on their farm. The discursive nature of FFS thus creates its own problem of perceived lack of discursive skills, a problem facilitators are supposed to deal with. But most facilitators did not have skills to encourage or motivate the marginalized illiterate learners to catch up with dominating literate learners. Often facilitators, just like school teachers, found it more convenient to move at the pace of the quick learners (i.e. local elites). This situation led to passive rather than active participation in which the "slow" felt less valued in the process, and chose to keep quiet or even to give up.

Debates help build reasoning capacity. Farmers' thinking and analytical skill were not adequately built up, as was intended by AESA. Elite elements tended to dominate speech, and this made some farmers shy away. Their discomfort contributed to drop outs. Dwindling numbers of farmers in FFS sessions reduced the richness of discussions. Debates, when held under adequate moderation, stimulate divergent thinking. Ideas produced make discussions richer, and with good facilitation the debates are shifted towards convergent thinking and inclusiveness. Simsons *et al.* (1999: 663) suggested that although members challenge and oppose one another, debate may sometimes encourage open-mindedness in decision making, hence foster comprehensiveness in decision-making. The listening skill in facilitators that would enable keeping track of all ideas in the debate or discussion was not adequately built. Listening goes with being observant. The two skills are necessary in understanding the dynamics among learners (farmers) and consequently choosing a more appropriate method to engage them.

Facilitators made FFS more formal than informal. They played teacher roles. In some FFS under CIP-IPPHM in western Kenya (Bungoma), marks were awarded to presentations from sub-groups, in addition to suggestions for improvement. Award of marks had two implications: one, this made farmers more active in observing, drawing and analyzing findings as well as increasing cohesion in sub-groups for good work (via a spirit of competition); two (and rather more negatively), it created an atmosphere that limited concentration on content/technology because of the pressure to present well, especially on subsequent sub-group presenters. A similar situation was observed in Uganda where a facilitator passed around with a red pen marking drawings made by (literate) farmers in their exercise books. The interest of the facilitator was more in how well the drawings were made than in the farmers' understanding behind the drawing. It also created tensions among illiterate farmers. Even the literates were divided up into the bright and the not so bright.

Discomfort with scheduling and activities in AESA. The practice to schedule AESA very early in the morning (FAO-IPPM, CIP-IPPHM) was difficult for most farmers to cope with. Normally, farmers went to their fields very early in the morning to avoid the midday sun. Farmers were more productive and concentrated in the morning in their home fields. In their perception it was better to turn to collective work (and thus FFS) after becoming exhausted in their own fields, since collective work required less energy, time and labour. Home fields were bigger and yet had to be managed with limited labour, therefore requiring more time. This partly explained why farmers, especially women, went late to the FFS training. A reason to schedule AESA in the morning was to enable farmers to see most of the insects in the field before they escaped, especially the nocturnal type. However, this did not work well, because morning sessions interrupted farmers' attention on home fields. FFS sessions were then rescheduled for afternoons after farmers had completed working their fields. Afternoon, to the farmers, meant after lunch. Local variations in times at which lunch was taken from translated into variations in the times at which different members reported to FFS sites. This was many times offered as the reason for facilitators being late in running FFS sessions (they claimed that farmers did not keep time). As a result, a pattern emerged of farmers waiting for their facilitators, sometimes for over an hour. In the absence of the facilitator, farmers did little or nothing on their own apart from conversing in small groups. This example shows how seemingly small details can have a considerable effect on participation. Impatient farmers walked away to attend to other activities. Some farmers (or groups) used facilitator lateness as a basis for their own absence, while others turned up only once they were sure the facilitator was present.

Searching for insects in the field was perceived as a game for children. It was difficult to catch jumping and flying insects. In some cases, some farmers felt obliged to come back to discussions with an insect, whatever trouble collecting it involved. AESA consumed a lot of time (3-4 hours) therefore kept farmers in the FFS for almost the whole afternoon, without any refreshment. This was more of a concern in FFS handling many treatments and experiments (e.g. CIP-IPPHM). Measuring leaf and plant heights and counting number of leaves and insects was mentioned as a tiring and often boring exercise by farmers. The number of plots to be sampled was often high. Each of the eight OFSP varieties were planted in three different ways (ridges, mounds and flats) and replicated twice. By the time farmers had sampled all the plots, they were exhausted for subsequent activities. To spend less time under the hot sun, farmers often made estimates. The data collection by farmers suited researchers more than farmers. Instead of understanding processes and outcomes, as expected in AESA, farmers spent almost all their time sampling, counting and recording data for statistical analysis. It may be worthwhile for farmers to understand the logic behind counting and measuring but doing it the researcher way strengthened the distance between researcher and farmer. This is the opposite of what FFS set out to achieve. Farmers often used estimations, and saw little value in being taken through the mathematics. A high number of experiments was more in researcher's interest (good statistical data sets might yield publishable results), but farmers were often left tired and confused.

Use of natural enemies of pests. Failure to see natural insects in action predating on insect pests, as taught by facilitators, minimized the enthusiasm of some farmers. "We want to see practically what insect (natural enemy) eats which insect and how, but we have never seen in reality the situation as taught…" The effect of beneficial insect on insect pests was not visible. It might have helped to create a more durable impression if FFS in Uganda had recourse to insect zoos, as Indonesian FFS (Fakih *et al.*, 2003). Failure of the facilitator to help farmers identify some insects as either pests or natural enemies discouraged some farmers. This suggested that some of the targeted insects were not important to the farmers (in the facilitator's perception).

From the way AESA was handled, we see FFS emphasised teaching and data collecting activity rather than either learning or skill building by farmers. Emphasis on discursive activity in AESA made FFS encourage the verbally strong, in most cases men. This situation minimized opportunities to exploit prevailing experience, or develop sound discussions and analysis upon which future points of action might be based. Few if any action-plans emerged out of the AESA. This could have been attributed to failure of the facilitators to internalize the process, as well as to the adoption of a convenient instructive mode of operation in which farmers were told what to do.

5.3.2 Energizers

During FFS sessions energizers (in the form of exercises, games, songs, poems, and stories) were used mainly to minimize boredom and tiredness, and to relax and keep participants focused, so that participants remained lively and engaged in the learning process. Energizers were mainly used by facilitators trained under FAO-IPPM, though not all of them took energizers seriously. Performing energizers to some extent minimized tension among participants, and increased informal interactions and cohesion of the group members. Some facilitators used energizers as an opportunity to enhance understanding of the topics discussed. This was through processing and reflecting up on the implication and applicability of the energizer in relation to either the issue being discussed and/or in relation to farmers' normal daily situation. Such energizers were well thought through prior to the day's meeting. Some facilitators especially under FAO-IPPM, A2N-ISPI (Busia) and CIP-IPPHM used energizers while others (mainly under MAK-IPM, MAK-SPUH and A2N-ISPI (Tororo) did not make much use of them. It needs some level of creativity and time to develop a repertoire. Other facilitators used energizers mainly to keep farmers active, but the themes bore no relationship with the day's topic. There were moments when the energizers did not help in refreshing the participants, especially when they were exhausted or lost. Some farmers, especially men, did not feel comfortable with some energizers. They associated energizers with children and did not take them seriously. Farmers were encouraged to bring their own energizers, and take over this aspect. Whether this was a sign of enthusiasm by facilitators to root FFS in local cultural idioms or evidence that energizers were neither internalized nor taken seriously as an activity is unclear.

5.3.3 Influence of facilitators on farmer interactions with curriculum

Facilitators' behaviour with regard to interaction with farmers had an influence on how farmers interacted with the facilitator and on the content of FFS. Some responses encouraged farmer engagement with the content, and there are those (by contrast) who acted in ways that limited such interactions, as will be discussed in this section.

Facilitator's behaviour that encouraged farmer interactions
Energetic participation with others in an activity builds confidence. Working for others, on the other hand, makes farmers more dependent. Discussions with FFS district coordinators and

program assistants suggested that farmer facilitators performed better than the extension staff in promoting farmer interactions and in engagement with the project and content. How the two types of facilitators related with the farmers largely explains this difference in assessment. Farmer facilitators were more informal, and gave more time to fellow farmers to discuss, try out and discover for themselves. This was particularly so in FAO-IPPM. Members in FFS groups under farmer facilitators were observed to be more confident and closer. Unlike in extension-led FFS, where some facilitators laid experiments for the farmers, following research recommendations, farmers in farmer-led FFS laid experiments by themselves. I witnessed this in one A2N-ISPI FFS in Busia where farmers did not know what treatments were in what plots (the study crop was groundnuts). Some plots were bigger than others, though the experiments looked very neat. "Our facilitator knows better because she is the one who demarcated the plots", said one of the farmers in the group. Perceived better performance of farmer facilitators could be attributed to factors below. However, that does not mean that all extension staff facilitators performed well and that all farmer facilitators were better. The factors to be looked at in analysing the general social distance between farmers and facilitators are discussed below. Most of the data in this sub-section are drawn mainly from FAO-IPPM and CIP-IPPHM where I had more interactions with farmer and extension staff facilitators.

Close and frequent interaction with the community. The farmer facilitators stayed within the same villages with the groups they facilitated. They met and participated in various community activities frequently. They understood the situation of fellow farmers better and therefore handled them with patience as colleagues or friends. Farmer facilitators were invested in local social networks and therefore shared social capital with other villagers. Extension staff stayed in town and only interacted with farmers during FFS sessions, which were once a week or fortnight. They were more visitors than residents in the villages. Some extension staff facilitators regularly checked on plots and farmers (especially those hosting and near the study site). Such groups were either along the way to the sub-county headquarters, near the sub-county headquarters, or ones with well-managed experiments. Groups near the sub-county headquarters were targets for visitors and were used as a tool for facilitators to prove that they were doing the work.

Use of simple language. Like most farmers, farmer facilitators were more comfortable in their local language, and were thus readily understood by other farmers. This helped minimise any inferiority complex due to failure to speak, write or understand English. Farmers did not only actively participate in, but understood what was being discussed during, the sessions. Choice to teach only what they were more comfortable with, and leave out what was difficult, made farmer facilitators more restricted, however, than extension staff who made efforts to teach all that was scheduled on the curriculum. Farmer facilitators used their own experience to sort out what to teach. They left out what they found difficult to understand during their time as learners in FFS. But using their experience as former FFS learners benefited their colleagues. This "editing" of the curriculum might be counted a strategy to avoid situations that would embarrass them before other farmers. Using the local language was a problem for most extension staff facilitators. Even in areas where facilitators and farmers spoke Ateso. Facilitators often reverted to English both in writing on the flip charts and in talking to farmers during sessions. This left most farmers feeling lost, and reduced opportunities for active participation and understanding by less educated villagers.

Involvement in many programs. Extension staff facilitators were involved in many programmes running in their district and respective sub-counties. Activities under different programs clashed during the season. Some were national programmes and had priority over other programmes or projects. These included activities of the National Agricultural Advisory Services (NAADS) in Uganda and the National Agriculture and Livestock Program (NALEP) in Kenya. Others activities were basically often viewed as opportunities for extension staff to earn some extra money to supplement their salaries. In any case, under each programme, the extension staff had to work with more several farmer groups scattered over different villages, some of which were distant. Farmer facilitators on the other hand hardly had more than one group to work with at a time. They therefore had more time for the FFS groups. A majority were more patient, and gave farmers more time to talk during the discussions, and to take part in the field related activities, unlike extension workers constantly rushing to catch up with other duties. This slower pace - perhaps to be taken as a sign of less self-confidence on the part of the farmer facilitator - was also a blessing, since it created more time for farmers to engage actively in learning freely.

Even so, some farmer facilitators did not spend enough time with the FFS groups. Farmers in Buyaya FFS (under FAO-IPPM) complained about the irregular appearances of their farmer facilitator. She came once in a while and met them as a group for about an hour at a time. Some farmers were happy not to waste valuable time, but others were dissatisfied with the facilitator coming irregularly and spending such a short time with them. "Even in that one hour that she stays with us, she teaches us in a rush, leaving most of us without (any) understanding". "I do not feel satisfied but I fear to tell her that I do not understand what she teaches", mentioned one lady. "...I feel we need more time to understand what she teaches us. Many times she leaves us lost..." These were concerns raised by members in the group. However, one woman mentioned that she felt happy when the facilitator failed to turn up, or when she spent little time with them because she found it a bother to attend in the first place. This emphasises that FFS was not really interesting to all who joined. The day the facilitator would meet the group, she would let the chairperson know so that she could mobilize her people. Much of the 'mishandling' of farmers in FFS was because there was no follow up of facilitators to find out how the farmers felt about the project and the assigned facilitator.

Prestige and status. But poor performance by farmer facilitators was the exception rather than the rule. A majority - especially under CIP-IPPHM and A2N-ISPI - took the FFS as a rare opportunity to attain some recognition for their communities. In villages which had earlier experienced FFS, alumni (and especially facilitators) were accorded respect due to their active participation in community work. A number of them had their villages at heart and made suggestions that would improve upon life in community during village meetings probably because they were more informed. The feeling of usefulness and desire to be seen as teachers in the community raised their interests in the task. This made them keen not to miss any session with their groups.

Looking at the interaction between farmers and facilitators, it comes out clearly that there is a need to support and facilitate conditions that reduce the social distance between farmers and facilitators, as a way of making technology itself more "approachable". Once the facilitator fits within the community socially, it becomes easier for farmers in the project to interact closely with the technology promoted. Fitting of the facilitator into a community, however, is more a matter of one's personal relations with others, commitment to work, and understanding of farmer lifestyles.

Facilitators' behaviour that made farmers disengage from FFS

Much as it is desired to see facilitators create a supportive learning environment for farmers in FFS, we also need to look into other factors that influence facilitators in implementing FFS activities besides their training needs. More attention needs to be put on understanding de-motivating factors, since these negatively influence interaction of facilitators with farmers. In the process of doing their FFS work, farmer facilitators and extension staff faced a number of circumstances lowering their morale. These included payment structure, lack of clear terms of reference, farmers' unreasonable expectations, heavy workload and lack of an incentive structure.

Payment structure. The payment system made some farmers uncomfortable. In FFS, all funds meant to be used by an FFS were deposited on the group's account. Expenditure was tied to four things: learning materials (stationery and experimental materials), field days, group development (income-generating activities for the group) and the facilitator's weekly allowance (fuel and lunch). A specific sum of money was allocated under each heading. An FFS group therefore was supposed to pay its facilitator when she/he taught them. Facilitators' fees covered about 20-30% of the total amount. Farmers did not comprehend reducing 'their' money by 20%. To them this was too much, especially since many facilitators also earned a salary from government. For projects where the researchers took time clearly to explain how the money was to be spent (especially during the inauguration period) - as was the case with those under the Makerere (MAK-IPM and MAK-SPUH) scheme - there was less resistance to paying facilitators. However, with projects like FAO-IPPM and CIP-IPPHM, that did not explicitly clarify the issue of funds to farmers, two situations arose. In some cases FFS groups were not willing to pay facilitators and sometimes facilitators demanded more than they were supposed to be paid. Provision of budget made the process more transparent since farmers then knew how much was to be spent on what and why. Some farmer groups stubbornly refused to pay facilitators, while others continuously complained that paying extension workers who also earned a salary from government was not fair.

Fear of failing to be paid after the work was accomplished lowered facilitators' willingness to commit to working closely with farmers. To most facilitators, it was discouraging to be paid by a mere group of farmer. They perceived farmers as persons supposed to depend on extension, and found it hard to conceive of extension depending on farmers, an attitude of looking at farmers as being inferior and not fit to act as bosses. In some situations it was difficult for farmers to part with the facilitator's money. They perceived all projects as government projects and could not see any reason why government should pay extension workers more than once. To them the facilitation fee was a second salary. "Farmers are not happy with money going to the facilitator because they feel the salaries we get as civil servants are enough". There were situations where for unclear reasons facilitators felt farmers simply did not like them or the technologies they promoted. To some groups, facilitators used their work with farmers only to meet their own selfish ends, in the name of projects, while doing nothing really to change the situation they encountered.

There was lack of consistent information about allowances for farmer facilitators. "Payment is not prompt and we are not sure about the amount to expect". Under the CIP-IPPHM, for instance, by the time the farmers returned from the facilitators' training, they were told that they would be paid 5,000 Uganda shillings (about 3 US dollars) per session/week. However, during the course of the season, the programme assistant mentioned pay of 3,000 Uganda shillings (less than 2 US dollars). This de-motivated the facilitators. In the case of Western Kenya, farmer facilitators received less (100-150 Kenya Shillings = 2,000-3,000 Uganda Shillings). "Why do we get less paid yet we all perform the same work as the extension staff?" asked one farmer facilitator.

The low payment by the projects appeared a critical issue. The rates were the same for all extension facilitators in all cases, yet some facilitators travelled longer distances and required more fuel and time. "Some FFS groups are located deep in the villages and are not easily accessible yet the transport allowance is a fixed rate (200 Kenya Shillings). We use more of our money to meet project objectives". Besides, the facilitators also had to attend to many impromptu activities, such as attending to FFS related visitors and attending meetings for which there was no funding. "We use our own personal resources to meet such requirements". One facilitator preferred that FFS projects continue as independent and de-linked from government activities. The implication is that for smooth operation of FFS independent facilitators are more adequate.

Farmers' expectations. In the beginning, farmers were interested and regularly attended FFS sessions and activities. However, with time, interest dropped. Intense poverty was one reason some were drawn to projects, expecting handouts and money. They then found they were less interested in the project's activities. Funerals obstructed regular attendance by some farmers, especially (1999-2003) when HIV/AIDS peaked along the Uganda-Kenya border.

The low uptake of the things taught in FFS made facilitators tense. Output of facilitation is measured in terms of how many farmers are carrying out the 'good' practices. "Trips and tours attracted people, but they rarely used the knowledge in their fields". This in itself implied that facilitators did not take time to understand why farmers did not take up practices observed during tours, or even reflect if use of tours was the most appropriate in those contexts. Sometimes unpredictable weather patterns - e.g. longer dry spells - destroyed the experiments and made it difficult for farmers to see any message facilitators were striving to pass across. Heavy work load was mentioned by two farmer facilitators in Soroti under CIP-IPPHM.

"There is too much work especially when it comes to field work" said one facilitator. "There are no health breaks, yet I have very serious ulcers" added the other facilitator.

Lack of incentive. Absence of appreciation from bosses made some facilitators feel that their services were not of value. There was hardly a mention of 'thank you' from bosses. As one facilitator under FAO-IPPM put it, "We only get a word of appreciation from our farmers and some visitors". According to the facilitators in Western Kenya with whom I interacted in a workshop, visitors from the district thanked farmers in the groups for the work, and did not mention anything about the work done by extension staff. "We do not seem to have any value, yet farmers are what they are because of our efforts and commitment" said one of the facilitators. An appreciative attitude motivates people to work because they realize that others value their contributions. There was a feeling that recognizing and encouraging very committed and hard working facilitators, even through provision of sponsorship for further studies, would encourage accountability.

This section on facilitators' engagement shows that facilitator motivation to work is one important area that needs attention. There is need to attend to factors that are likely to kill morale or motivate facilitators. The most critical issues cluster (as ever!) around money. A clear payment structure, and a timely mode of payment, via which facilitators receive their allowances without difficulty, would be highly motivational. The other very important, yet often ignored, issue is about incentives to good work. A word of appreciation costs so little but does so much to motivate people.

5.3.4 Fund management in FFS

Funding in FFS groups was tagged to four main activities. These activities included purchase of learning materials (for AESA), field days, group development and facilitation (Figure 7).

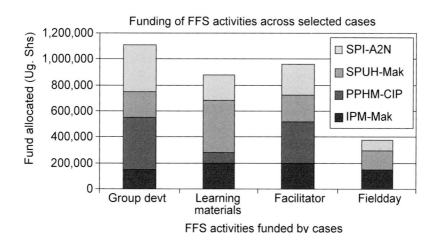

Figure 7: Allocation of funds to FFS activities.

Each FFS group was allocated a maximum of US$ 500 for a year but in batches. The money was deposited on group accounts. Money provided an incentive to farmers to work together. However, lapses in arrangements for accounting for these funds at group level seemed to encourage its misuse.

Handling of funds allocated to FFS activities

FFS groups opened bank accounts on which funds to run FFS were deposited. Only the first FAO-IPPM groups received cash directly. The chair person, secretary and treasurer of the group were signatories to the account. The idea was to make farmers fully in charge of how to use the money for the good of the group. However, this was abused in some circumstances in which signatories withdrew all the money without the knowledge of the rest of the group and divided it amongst themselves. To minimize misuse of group funds by a few individuals, FAO-IPPM FFS, ISPI-A2N introduced some conditions. One included having the facilitator as an external signatory. The other was written evidence of consent (minuted) from the group members, including their names and signatures as back-up for withdrawal from any FFS group account. Although most of the funding was provided as a grant, A2N funded groups worked on a semi grant scheme for group development activity, where farmer groups were expected to refund 50% of the money they had been advanced. The semi grant system failed, however, since the farmers treated it as a gift, and saw no reason to pay back the agreed amount.

Learning materials (input) allocation and purchase. The MAK-SPUH project allocated considerable amounts to learning materials because the activity required protective gear, spray pumps, pesticides (which were expensive) and improved seed varieties. CIP-IPPHM allocated the least amount to learning materials because the programme assistant purchased sweet potato vines of different varieties and delivered most of them to the groups, including note books and pencils to the farmers. Under FAO-IPPM, learning materials were purchased by the coordinating office (FFS secretariat) and sometimes by facilitators, given that most FFS groups were deep in the villages and transport costs were high. Stationery was bought in bulk from the city (Kampala) where it was reportedly cheaper and then re-distributed to the FFS groups. However, this arrangement was abused by some facilitators, who bought cheaper items of lower quality, with quantities acquired not always matching the amount of the money handed over. Sometimes some inputs were supplied too late in the season.

Purchase of inputs, especially in farmer-run FFS, was undertaken by the secretariat. According to farmers in the Bakisa FFS group (in Busia) there was no transparency and accountability. The costs that the facilitators attached to the inputs were so high that farmers never believed they were a correct reflection of expenditure. Farmers did not know how much of 'their money' went into buying the inputs. Some of the inputs were not used. Flip charts remained unused and farmers interpreted this as a waste of their money through buying what they did not need. In another instance, fertilizers like DAP, Urea, and Single Super Phosphate (SSP) were stockpiled and never used. Farmers knew the inputs were bought for them, neither they or their (farmer) facilitator knew what they were, or how to use them! This probably explained why farmers in this group felt that the project imposed activities on them. None made effort to find out what the inputs were for - an indication of very low interest in the inputs. Besides, farmers in this group did not agree that their soils were of poor fertility. They kept insisting that their soils were fertile enough and did not require application of fertilizers. "Fertilizers spoil the soil and we do not want to be like Kenya", they said. Kenya's soils are poor and farmers often use fertilizers. This suggests a misunderstood correlation between fertilizer and soil impoverishment. Farmers preferred that the secretariat and/or facilitator would tell them what they were planning to offer/bring, prior to acting upon it, in order to gain agreement and consensus. In the case of the wasted inputs, it would have been better for farmers to be trained first, so that they understand what to do with the inputs.

Some FFS did not get all the expected money, but received no explanation. It was mentioned that at that time, 2002, the secretariat received a list of more FFS than money had been released at FAO headquarters, in Kampala. Not to discourage any of them, the decision was to divide up the available funds amongst all the registered FFS. That was why instead of the expected amount (500,000 Uganda shillings) FFS groups all received less. But lacking information about this change, the group assumed the executive had embezzled an unknown balance. The perception was that the chairperson, who often went to the district head quarters, took the money and shared it among some members on the executive. Lack of clarity about the situation with the remaining money created some level of mistrust among the FFS group members, which in turn affected the quality of their interactions and hence their ability to learn. There is need to update the farmers in the FFS in question with any information concerning their field school activities, in order to minimize chances of misunderstandings jeopardizing otherwise well intentioned project activity.

Facilitator's allowance. There was some variation in the facilitators' allowance due to the difference in number of sessions and duration in the field. The sweet potato crop spent longer in the field and therefore the facilitator had more sessions with the farmers. Farmer facilitators, however, received less money for their allowance than their counterparts, the extension staff. The reason given was that since farmers stayed with and belonged in the same communities as the FFS groups they did not move longer distances needing fuel. They were given 30-50% of the extension staff allowance depending on the budget of a project. In some cases, farmers manipulated and did not pay the facilitators. Sometimes facilitators manipulated farmers and took the money without offering a service.

Group development fund. A special fund was given to a group to encourage members to manage a collective income-generating enterprise. A2N-ISPI provided more funding for group development compared to other activities because it was intended to serve as a loan to be paid back after a year. In other projects money was provided as a grant. The kind of enterprises set up varied from group to group. Group leaders, and sometimes facilitators, influenced the choice of enterprise. Such advice was not always for the benefit of group. In Kiboga, leaders of one group preferred buying a mobile phone as their enterprise. There was objection to this idea when members failed to agree on who would keep the phone, to avail it to all members at all times, ensure credit for use and take the trouble to keep it charged throughout. There was no electricity in the village and charging the phone meant making frequent travels to the district head quarters - an insupportable expense. In Mukono, leaders of one FFS group decided to buy a second spray pump to increase chances of each member accessing a pump with minimal delay. This divided the group. Some members did not like the idea while others supported it. However, it ended up that the second pump was used by friends of the leaders, who neither belonged to the group nor paid for the privilege of using the pump.

Most groups opened up areas of land for groundnut and maize production as their commercial enterprise. Others bought a few female goats - 4 to 6, depending on prevailing market price - divided the members into sub-groups equal to the number of goats and gave each sub-group one goat. It was up to the sub-groups to care for its goat. Each time the goat kidded, the kid was passed to another member. The cycle continued until every member had a goat. Thereafter, the next kids belonged to the group and were sold off to generate income to solve any problems that needed money. Some groups just decided to divide the money amongst the members. Others used the money as capital to set up a fund for loans to members.

In addition to a group account, every FFS group was encouraged to have a visitor's book in which all visitors to the group, irrespective of level and where they came from, signed and made remarks about what they saw. The remarks were, overall, positive. Farmers made little use of the information in the visitors' book. Visitors who signed might be used as a link in accessing other services. Sometimes farmers used the remarks for advice. Facilitators sometimes abused the visitors' book. They signed the book without attending to the group.

Provision of money to ensure smooth running of FFS activities shows commitment to build local capacity. It also encouraged establishment of commercial enterprises to boost individual incomes through group activities. Absence of accountability regarding such funds (or grants) at group level encouraged some individuals to take advantage and exploit the group. The arrangement was subject to abuse by some individuals, who looked at projects as sources of income to fund personal interests but not as an opportunity to boost development of the community at large. A better way might be to establish a matching grant scheme, where a group is given a sum of money equal to what it contributes as a group, therefore ensuring that every group member makes a contribution. In that way, every body becomes vigilant.

Revolving fund scheme to sustain the FFS model

Towards the last year of operation (June 2002), a revolving fund scheme was established under the FAO-IPPM project. The idea was to make the system self-financing, and therefore to mobilize funds to benefit more farmers continuously. The money was then given as a loan with 100% recovery in a period of about eight months. Facilitators identified five groups (one staff run and four farmer run) in every sub-county. The recovered money was then to be advanced to other groups. All FFS groups under this regime were supposed to cultivate vegetables, a crop expected to have quick returns. The staff-run FFS received a loan of 1,190,000 Ugandan Shillings while the farmer run scheme received 780,000 Uganda Shillings. The difference reflected the difference in levels of allowance paid to the two categories of facilitators. Whereas extension staff facilitators were paid 10,000 shillings per week, the farmer-facilitators earned 3,000 shillings.

Three months after giving out the loans, FFS alumni were encouraged to form networks at sub-county and district level to manage FFS groups under the revolving fund, identify the new benefiting groups and ensure recovery of the funds. Sensitizing farmers about IPPM and lobbying for funding were other roles to be played by the network. Every FFS group elected one member to represent it at the sub-county. Most of the representatives were chairpersons of groups who were also farmer facilitators. Out of the various group representatives at sub-county level, an executive board of eight members was established, consisting of a president, vice-president, secretary, treasurer and four ordinary members. Each sub-county in turn chose one representative at the district level. A district executive was elected from members of the sub-county executives, with an attempt to have all sub-counties represented at district level.

The FFS network structure soon ceased to be functional mainly due to lack of support. Members were expected to work on a voluntary basis, which proved impractical in Ugandan conditions. There was no adequate preparation of either farmers' groups or facilitators about the transformation of operations from grant to loan. A mechanism to recover the loan was not worked out prior to giving out the loans to farmer groups. This partly explained why the network failed to recover the loan for the revolving fund. In a review meeting held on 29th April 2003, with the objective of coming up with a collective format for report writing, members of the networks registered difficulty in recovering the Revolving Fund.

Reasons why the revolving fund scheme did not take off
The failure of the network to execute its duties involved many reasons, some of which are discussed below.

Misguidance by local leaders. Some local politicians informed groups not to refund any money, by suggesting that the money was meant to be a grant. These politicians were representatives of farmers at various levels, and farmers thus had a reason to consider this reliable information. There was a case where some FFS were advised by their politicians to cheat the facilitator by agreeing on a different time for meeting, and not showing up when the facilitator expected them. One group told the facilitator that they would be meeting at 8:00 am, but whenever the facilitator turned up he found them returning home. "We cannot pay you since you never honoured the time we agreed to meet", they told the facilitator, who later gave up.

Disputes over refunds and costs. FFS groups were not willing to refund 100% of the initial investment because most of it was paid to the facilitators or was spent on buying experimental materials - at over-valued prices, in their estimation. In some cases materials such as flip charts and crayons were never used in teaching farmers in FFS activities. Instead, facilitators used them to train other farmers under the NAADS program.

Group disintegration. Some groups disintegrated after receiving the loan and it was difficult for the network members to trace the group members. "Every member we associated with in such groups denied belonging", mentioned one farmer facilitator. When farmers heard about money, they rushed to form or join groups without taking the trouble to find out who was in the group or why. Need for credit caused all this. Such groups split immediately after getting the money, whether all or few members benefited. In one case, some villagers joined a group without knowing its objectives. Some people find themselves used by others to meet private ends in the name of group membership. This shows the hazards of group formation when induced by project incentives. With some exceptions, it seems difficult to assess beforehand what individual motives impel decisions to join an FFS group. The ideal is to work with interested, committed and serious group members, but commitment is co-dependent on group solidarity, and solidarity fails to form where opportunism rules.

Lack of collective commercial activity. Some groups divided up the money among members and had little or no collective activity on the ground. Most groups divided up part of the money (meant for commercial plots) as capital for petty business. There were situations where the chairmen of FFS collaborated with secretaries, to withdrew all the money and divide it equally between them. When one case was discovered the miscreants promised to pay it back from the proceeds of trans-border trade but getting hold of them was not easy.

Unclear communication. In some situations, the FFS groups demanded an official document that stipulated the FFS network was obliged to collect the money, but this had no practical effect. By the time farmer groups were being given the loan, there was no longer any idea about a network! Sub-county administration was not supportive because they felt it was not their responsibility. The capacity of these farmers to deal with or handle the loans efficiently seemed low and needed to be built before giving them the money. Besides, the communication on how to use the money was not clear to all.

Lack of interest and commitment of network members. Network members who were once cheated by other farmers in FFS were reluctant to engage in a recovery exercise. They felt they would never benefit from the money. Network members suggested a way forward in ensuring recovery of the loan. It was suggested to include a letter written by the District Agricultural Officer through the Chief Administrative Officer to the sub-county chiefs. Other ideas were to trace and use documents signed between farmer organization and the programme for commitment to the formation of a recovery task force at sub-county level with representation from all groups that had benefited from the loan, including regular updates about recovery of loans. None of these suggestions was pushed forward partly because no one was given the responsibility to ensure implementation and partly because there was no incentive to motivate the network members to do the work. A2N-ISPI project 'borrowed a leaf' from the loan scheme but recovery was targeted on the group development fund. Using the potholes faced by FAO-IPPM as lessons, A2N-ISPI project prepared group leaders of selected participating groups and facilitators in business management and even discouraged the FFS groups from investing in farming activities. A different network, but with some members from the old FAO-IPPM network, was formed with a separate account.

Unrealistic terms. Farmers were advised not to invest in non-crop related commercial activities because presumably this would lock the loan capital up in the enterprise for lengthy periods of time. But there was reluctance to investing in crop enterprises, as required by the program. At the time, there was a long dry spell followed by heavy rains undermining the viability of many farming ventures. Groups preferred to invest in brick making, piggery, fish farming and poultry management, all of which needed more than a year to recover the loan.

Failure of the revolving fund scheme to work was mainly attributed to an inadequate and rushed way of doing things. There is need to work out a comprehensive strategy to run such schemes to ensure sustainable objectives. One of the most critical issues is thorough preparation of farmers, local leaders and facilitators on how to access money. Consensus and clarity on roles and responsibilities in the operation of revolving fund schemes prior to implementation would safeguard functionality. Providing the loan on supply (through convincing a group to take, part as was done by local leaders during mobilisation) rather than on demand (through competition for funds according to the merit of the application) encourages farmers not to take repayment seriously.

5.3.5 General perceptions and contributions attributed to the FFS model

Considering the internal organisation of FFS, we arrive at the following points. Firstly, the process and objective of AESA was not internalised or clearly understood by farmers and facilitators. Implementing and guiding AESA was based on what facilitators understood it to be, namely quantitative data collection, instead of developing an understanding of processes and interactions leading to an observed outcome. This means that much emphasis in any future application has to be put on understanding of the AESA process as an aspect of the wider objectives of FFS. Secondly, farmers and facilitators need an environment that promotes mutual thrust. Project incentives to create such an environment through group formation often turn out to have the opposite effect. This is clearly apparent in the issue of management of funds. Providing funds to the group builds local farmer capacity in managing finance but providing funds as a grant, without strict accountability requirements, exposes FFS to abuse by people who see it as a source of easy funds. Third, smooth operation of a revolving fund scheme requires collective effort and clear roles to be assigned to the players involved, especially the direct beneficiaries.

5.4 Integration of lower with higher structures

Whereas the section above (5.3) looked at the internal interactions, this section (5.4) looks at the interaction and hence integration of FFS (the internal) with other actors outside FFS (the external). The integration was through three main activities of monitoring and reporting, field days, and field tours/visits as discussed below.

5.4.1 Monitoring and reporting of FFS activities

Monitoring of FFS comprised three elements. One was the regular AESA carried out by farmers on a weekly basis. This was mainly on pests and general crop performance involving pictorial presentation on flip charts. On a day-to-day basis, host farmers monitored experimental plots and often reported verbally to other farmers or the facilitator in cases where there was a problem with a strange pest or disease (examples are the case of leaf miner on groundnuts in Busia and Tororo, under A2N-ISPI, and wilt in Kales in Busia, under FAO-IPPM). Facilitators also monitored technology performance. A second way was through the activity of a program assistant, but that was not a regular occurrence, due to the huge work load of these assistants. As a result, monitoring became a task for the facilitators. The third option was the use of external consultants, as in the case of the FAO project. Consultants (external project evaluators) provided an outsider's point of view, but a problem with this was that a task usually taking more than a year or two might be assessed in less than a month. The chances of missing out processes crucial in explaining outcomes seem quite high. Furthermore, terms of reference often guaranteed that the exercise produced results mainly interesting to scientist and funding agencies. For improvement of FFS on a daily basis there is need for a robust regular internal monitoring system.

Reporting tended to assume a one-way hierarchical arrangement from local level to donor level. The higher the report reached the more abstract it became. Through AESA observations and discussions, farmers reported to each other about performance of experiments both verbally and in pictorial form. Facilitators analyzed and provided reports, either in formal meetings (workshop or retreat) organized by researchers at international level in collaboration with actors at lower levels, or (mainly on a verbal level) to programme assistants. It was rare for facilitators to mention what was not working in front of their bosses, as this tended to get them tagged as non-performers. What people said or wrote in reports depended on how free they were with their bosses and more whether the bosses encouraged them to speak freely without implication. A learning-oriented boss would tend to get more, and more varied, information from below. After sorting and analysis, program assistants provided regular reports (at least quarterly) to project leaders at international level who presented a final technical report to donors. However, such reporting was more on outcomes of treatments that performed better during the experimentation. There was very minimal, if any, reporting on process, and process difficulties, or on 'failed' experiments. Whether or not success was the only message the higher levels wanted to hear this is what the lower levels mostly assumed and loyally provided. The problem is generic within development work and science. Despite the philosophical argument that science proceeds by falsifying hypothesis there is little practical analysis of "failed" experimentation, except in the sense of a "post mortem" (finding reasons to terminate a project or condemn a failure).

The final reporting done by technical people often ignored the AESA reports of farmers, and gave no really detailed picture of what practically happened on the ground. As noted, even some farmers assumed that observation was simply a frill to be dispensed with. Sharing of final reports varied across projects. Documentation of formal project meetings was done by a member on the international team. This created opportunity for the report to contain what the researcher took as important in his/her case, leaving out other issues of possible high interest to others. Under FAO, such reports were not shared with facilitators and other participants present. A certain house style to these reports reflected the interests of FAO and funding agencies. This reinforced a picture in which lower level actors are accountable to higher level actors but the reverse is not considered to be the case. Mutual accountability creates transparency, commitment and responsibility on either side. In the case of CIP-IPPHM project, reports or proceedings were regularly shared with all people who participated in various workshops, including myself, in hard copy format. For projects that did not have programme assistants, like those organised by Makerere, the facilitators provided both verbal and written reports to the researchers who compiled a report to the funding agencies. FFS projects organised by Makerere included post graduate training at masters' level, and under this arrangement M.Sc. students helped compile reports. Such reports tended to be more suitable for the needs of researcher and funding agencies (mainly for accountability purposes) than of benefit to farmers and facilitators. More attention to the process of how farmers actually used technology to improve upon their farming system would have been useful. When facilitators reported they mainly focused on presence of a bank account, certificate of registration with the district, group size with number of men and women, enterprises, yield differences, and how much money was used for which activities.

Under FAO-IPPM there was a difference in reporting between Kenya and Uganda. In Uganda, there was hardly any written reporting about project activities. Reporting was more in the form of facilitators sharing experiences informally in regional review meetings prepared by FAO. According to facilitators in Uganda, there was no written report, because they did not know what to write about in the reports, a confession made during a one-day FFS review workshop I co-facilitated in Soroti. The reporting culture, especially in government institutions is weak. In Kenya, by contrast, there was a format that facilitators completed on a monthly basis. These forms, taken as reports, were handed over to the FFS coordinator, who presented them to the programme assistant. Completing the forms with the name of the FFS, its location, membership (numbers of men and women), enterprises attempted, month schedule of activities, and remarks, seemed to have been mainly a bureaucratic routine, because little of this information, so far as could be discovered, actually fed into any active process for subsequent improvements. The recommendations or suggestions for improved operation or functioning of the following FFS (the next season or year) resulting from previous experience ended on paper. There were situations where the same complaints kept being raised month after month, with the implication that nothing were being done. Reports were nicely filled artefacts but not instrumental documents. Although formats saved time in reporting and made it convenient for the scientist to analyse data, they tended to obscure or exclude any reporting about processes and surprises. Interaction of farmers with the technology is in practice full of surprises, and this is where the most valuable lessons are to be learnt.

Documentation and monitoring & evaluation (M&E) of FFS

The information written down in FFS documentation focused more on scientists and funding agencies, and tended not to circulate. Reports provided or written by facilitators were kept by scientists. Field workers and farmers did not have chance to look at final reports, even though such results would have been highly useful if fed back into the system for improvement of the project in general. Sharing findings with actors at the lower levels (facilitators and farmers especially) would encourage a culture of more objective monitoring, evaluation and documentation, reflecting the realities on the ground. There was little sense (as in science more generally) that FFS needed to investigate its own procedures on an ongoing basis (i.e. there was little or no in-house notion of "technography"). Assuming the overall institutional culture of development aid to be dominated by "impression management" facilitators often perceived themselves as in danger of being judged by the process, and so tended to report what was "palatable" to the scientists, whether or not this was actually what was going on. After all, scientists were not on the ground, and appeared only once in a while, especially when accompanying visitors to the project sites or during field days. On these occasions time was too limited to get the true story behind what was seen and heard. The information needs of players in FFS were not clarified, and therefore the roles played by each actor in collecting information tended to be based on do-it-yourself or opportunist assumptions.

This would suggest there was need for a clearer monitoring framework to guide data collection, evaluation, reporting and information use. Sharing reports including sources of information would encourage more objective standards of monitoring, evaluation, and reporting and documentation. Ideally, monitoring provides information that feeds modified action with the aim of making improvements. However, unless the results of monitoring are well captured and kept in a written form, there is no reference point.

Monitoring a process is important, given that one is in a better position to interpret or understand and learn from the process and plan new outcome more effectively. Collected information is useful only when put to use, which comes with reflection to pick out action points for subsequent activities on the ground. A study in Peru by Groeneweg and Tafur (2003: 301) revealed very limited use of monitoring and evaluation (M&E) results in FFS mainly due to reliance on analysis done by outsiders. Provision of a report writing format, as was the case of FAO-IPPM in Kenya, was perhaps even more limiting in terms of failure to allow for processes and surprises. Facilitators did not know the value of documenting the entire processes, interactions and activities as they unfolded in FFS Probably an open structure to encourage creativity and guide report writing would be adequate. 'Dry' reports rarely capture an audience's interest. But who is the targeted audience in this context? Details about what, where, when, who, how, and why with reference to what occurred, provide 'life' to a text. This even richer if it flows in a cycle of reflection on action, cause, and effect, as in "action research", where the interaction of action and reflection are emphasised.

Action and reflection are important in producing practical knowledge directly useful to farmers. The practice of AESA was supposed to provide systematic development of knowledge about process and outcome. Action without effective reflection and understanding is as blind as theory without action is meaningless (Reason and Bradbury, 2001:2). What farmers did and said with reference to the chosen technologies was rarely catered for in the report. The emphasis was on technology performance in terms of maturity period, resistance to pests and yields. Farmers' voices, commonly excluded, are essential in the final or technical report. This is because they are the 'marrow' of the project and give direction to how fitting a given technology is judged to be in local contexts. What farmers perceive and need is different from what facilitators, researchers and other stakeholders require.

In monitoring and reporting, emphasis was put on performance of technology. Sometimes it was about what was expected to have happened and not what exactly happened for fear of being seen as not performing. Ignoring society in technology use is false economy in that it obscures any clear picture of the relevance of the technology or even how to improve it to suit societal needs. To minimise bias in reporting skewed towards technology, it was important to look at interactions between farmers, teaching method and content/technology as illustrated (Figure 8). Information on interaction, among farmers, between farmers and content, between farmers and facilitators, between teaching method and content, and between farmer and teaching method would provide better input to revision of methodology and content of FFS. In FFS, reporting was linear and one direction. Instead of a linear one-way mode of reporting from facilitator to researcher or funding agency, a cyclic and multi-directional mode would be preferable, with stress on the mutual accountability of all stakeholders (Figure 9).

Because farmers, facilitators and researchers feared blame for reporting what was happening, they preferred to keep quiet. Farmers would make statements like "I will not tell you because you will end up asking our facilitator and he will get annoyed with me or the group..." as I kept inquiring about interaction between farmers, facilitator and content. The culture of blame "kills" many things that would have been otherwise rectified. Instead of blame, monitoring

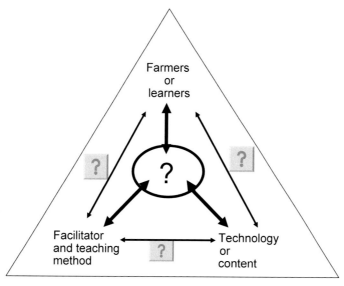

Figure 8: Cornerstones for improved reporting.

Figure 9: Multi-directional reporting cycle.

and evaluation processes offer opportunities to identify areas of improvement for the good of the projects in designing more feasible technologies and curriculum. This would minimise reporting and sharing of "false" information with regard to interaction of the farmers, technology and curriculum. Even in the reporting made by facilitators and others in the FFS, everything always looked perfect, as if the authors had been warned against mentioning any challenges being faced! Changing what amounts to a culture of deceit and blame is a wider challenge for all development project activity and not restricted to FFS alone.

5.4.2 Field days

Farmers in FFS shared with the rest of the community what they had been exposed to during FFS training. This was mainly done through showing and explaining what they did in the plots. This may not contribute much to farmer-to-farmer knowledge sharing, given the need for hands-on training processes. In spite of increased knowledge linked to increased productivity reported for some African countries, e.g. Kenya (Mwagi et al., 2003; Bunyata et al., 2005) and Zimbabwe (Mutandwa and Mpangwa, 2004), studies in Asian countries, specifically Indonesia (Feder et al., 2004) and Sri-Lanka (Tripp et al., 2005) revealed no effective diffusion of knowledge from FFS farmers to non-FFS farmers. This may be partly attributed to limited interaction between FFS farmers and the rest of the community, much as Rola et al. (2002) found out for the Philippines. This section does not attempt to measure retention of knowledge, but seeks to demonstrate the point that field days did little to enhance sharing of knowledge between FFS and non FFS farmers, nor in truth were they intended to. They were more a kind of ritual display of development potential enacted before a high-status audience of political actors (as described for a British- funded participatory agro-technology development project in India studied by Mosse (2005))

Across all projects, attention was paid in field day arrangements to guests from higher social strata. The seating arrangement, with farmers on one side and visitors on the other, the limited time devoted to FFS farmers explaining what they learnt (maximum one hour), and the allocation of most of the day to formal speeches from invited guests from higher levels simply did not allow for real interaction between FFS and non FFS farmers. Priority was given to the 'big' bosses who moved ahead of the rest of the farming community to the FFS fields, which were often at quite a distance. It is a telling commentary on the real purpose of these events that unlike the farmers, guests from district and national level were provided with transport. A tendency to join field days with other functions such as World Food Day pulled in many people but overshadowed the intentions of the field day.

Field days were chosen by facilitators who targeted very attractive growth stages and appearance of the experimental plots (i.e. with the approach of the harvest). Poorly performing plots were not chosen as field day sites because "there was nothing to show", according to the facilitators and a fear of not being seen to be serious by the bosses. Again, there was no space or time to discuss "negative" lessons. Local political leaders presided over FFS functions. In situations where all fields were 'bad', as judged by the facilitator, the FFS did not hold a field day. This was more pronounced in FFS under FAO-IPPM in Western Kenya. For example in 2003, Butula division failed to hold a single field day. One report is frank about the thinking "all fields were miserable because of drought and it will be a shame to show such (fields)..." Field days were more of a political activity to show highly placed people at the district and national level to "enact" the successful working of FFS projects. Dodging 'bad' fields was in a way dodging negative criticisms from politicians.

FFS training sessions held after field days registered less turn out of farmers and in some FFS farmers simply never turned up to finish the curriculum. This was observed under CIP-IPPHM. Facilitators made efforts to be present but found the agreed meeting place - a classroom - was empty. However, not to be accused of absconding, in an FFS where members had to sign their visitor's book each time they turned up, the case was covered by the remark "no farmer present". One said "after the field day what else do we have to learn?" A field day thus was perceived as the end of FFS training. It appeared to be a sign of relief, an implication that farmers were burdened by the frequent/regular FFS training and activities. But it also offered a moment of ritual closure. Like at a funeral, a life well lived had been celebrated, and now the deceased was to be allowed to return quietly to the world of the spirits.

5.4.3 Field tours and visits

Tours and visits raised FFS farmers' awareness about existence and feasibility of other technologies outside FFS in their villages or districts. Tours and visits were organized by programme assistants and facilitators. Farmers toured research institutes that hosted technologies promoted in FFS. Farmers in CIP-IPPHM-FFS had a tour to the National Agriculture Research Organizations (NARO) that hosted the sweet potato program. They visited fellow farmers within and outside their districts. Some even visited farmers in other neighbouring countries. For example farmers in Western Kenya visited groups in Uganda and in the Coastal Region of Kenya, mainly under FAO. Those in Uganda visited groups in other sub-counties within the same district, especially under CIP-IPPHM. Besides socializing, some farmers used this as an opportunity to try out what they saw farmers doing elsewhere. Some farmers in Busia-Kenya picked up (local) poultry enterprise as a commercial activity seen on their visit to the Kenya coast. In a similar way one farmer picked up banana management practices (mainly spacing, mulching) observed in Uganda. Out of curiosity, while on tour in Uganda on invitation of the research institute (NARO), one farmer picked some vines of an unknown improved sweet potato variety, and multiplied and planted them at his home. The varieties were under trial and had not yet been released for on-farm evaluations. Farmers had been cautioned not to pick anything from the experiments. He was the only one with the new variety in his village. The variety has still not been released for on-farm evaluations but it remains with this farmer and a few friends. Researchers were astonished to see a variety they had not released already doing well in the field. This shows that it is not always necessary to test technologies extensively before they find their own route to successful application. It also suggests that the farmer-led experimental elements in FFS might be a better route to adoption than a research-led analytical approach.

Farmers within the same group also visited each other's fields and made suggestions for improvement to the host. This was observed in one FFS group under CIP-IPPHM in Bungoma, Western Kenya. However, the suggestions were not taken up because farmers did not feel comfortable about being directed by peers about what to do on the home farm. Some of the advice given ignored the socio-economic situation of the home in question. Group members rarely took time to find out why their colleagues did what they did. As a result some insulted others as lazy, a situation that minimized further interactions. In Bongoma, one farmer was not interested in hosting his colleagues because he never grew sweet potatoes and the group was a potato group. He just wanted to associate with the group to win their support when needed, not because he was interested in the technology.

Taking farmers on tours was a very expensive activity given the logistics involved. For this reason projects did not organize tours for farmers. Tours were mainly organized by FAO-IPPM. CIP-IPPHM encouraged visiting farmer groups within the same village or sub-county because it was cheaper. To most facilitators, trips reinforced application of knowledge gained. To many farmers trips were recreational and social. They rarely got an opportunity to travel outside their districts normally. All were interested in the tours. All costs were paid by the project.

In this subsection we have seen FFS integrating farmers or communities with structures at higher levels through reports, field days and tours. The integration process was unidirectional and not two-way, with lower levels accountable to higher levels The emphasis of the reporting system on the technical content translated into an accountability to researchers and funders. The field days turned out to be activities where project implementers sought back up from higher authorities, to justify existence and funding. Field tours and visits were perceived differently by the two levels. The aim at the higher level was to expose farmers to other technologies while farmers seized on trips to broaden their social horizons. When using different methods to integrate lower levels and higher levels there is need for a clear objective and strategy to reshape FFS into a participatory model serving interests or concerns at both levels.

5.5 Linkage to other local activities or structures

FFS, like any community development programme, has all sorts of wider effects. Presence of these wider positive effects helps justify project investment. This section explores community involvement in FFS projects and linkage with other local activities.

5.5.1 Other information sources and complementarities with FFS

Before FFS projects began in the different districts, farmers accessed agricultural information and technologies from a variety of sources, including fellow farmers (individuals and groups), local council leaders, church groups, researchers, government extension workers, and non-governmental organizations, including community based organizations (CBOs). Methods through which farmers acquired information included lectures, reading material, short courses and workshops, media (radio and newspapers), demonstrations and experimental plots, and field visits and tours. Sources often complemented each other given that different farmers were exposed to different information opportunities, and that these varied from district to district.

A study by Riesenberg (1989), suggests that farmers preferred more interpersonal methods, especially demonstrations, experiments, tours and trips of the kind used in FFS. In North-eastern Uganda, farmers preferred seminars in addition to demonstration and visits/tours (Turrall et al., 2002). The interactive and stimulating nature of methods enhanced formation of informal networks through which farmers shared and acquired information. As Conley and Udry (2001) found out in Ghana, information regarding farming flows through a relatively sparse social network. The major information sought or acquired was mainly in crop production/husbandry and very little exposure to animal husbandry, marketing and post harvest handling.

FFS was not fundamentally different from other projects or NGOs in the way it recruited members and ran its groups. FFS alumni justified their involvement retrospectively in terms of FFS offering the best information. Promotion of interactions through teaching methods used, especially AESA and field days, were seen as factors underpinning FFS superiority as an information source. As discussed earlier (chapter three), most groups under FFS were formerly set up by various organisations (with donor money and self initiatives). The decision of these groups to join FFS (and therefore get referred to as FFS groups) had no links with members of the groups reviewing their groups, other groups or methodologies used but for a mix of reasons. Comparison of information sources, however, was triggered by my inquiries into other sources of agricultural information; how the various sources were rated and reasons (criteria) underlying the ranks assigned to specific information sources. The sample size of each FFS group varied. Members (in the range of 9-16) who turned up presumably represented the group. Members of each FFS first mentioned the various sources of information then each member ranked the sources. Results were aggregated and average ranking taken for the entire group. The reasons farmers raised in a group discussion to justify the average ranking assigned served as the criteria (with which farmers compared FFS with other organisations that offered similar services). Although in practice farmers shopped around to some extent and managed to extract some complementarities from the mix of services available to farmers (see comparison tables in Annex 4) the system must be described as inherently disorganised. Whereas some agencies worked with existing groups formed by other organizations, others preferred to make their own groups, most of whose members belonged to older groups. In operation, every organization went its own distinctive way and few if any ever deliberately sought to strengthen or build on what others did. This, in many situations, led to duplication of services that fatigued farmers. In one case in Busia, for example, a poultry management project by NAADS was preceded by a similar project from Sihubira Farmers' Ogranisation (SFO), with the same groups. A request by the NAADS hired service provider (SFO), to build on what was already there (had been done by SFO before contraction to provide services under NAADS) was not heeded. The real service provider was an FFS alumnus. FFS, like other agencies, saw its role primarily as to diversify the service provision landscape, and fill a gap not being filled by others. The focus was more on elaborating the internal vision of experience-based learning about technology, and not on integration within a matrix of service providers. The projects examined had no specific methodology (e.g. technographic survey) for "mapping the landscape" of service provision to see what complementarities it offered.

5.5.2 Integration of FFS in NAADS

Some extension staff (former FFS facilitators) integrated experimentation and AESA while teaching farmer groups under the NAADS program. Under NAADS, public extension workers were not allowed to provide services to the community. Service provision was considered to be work for private service providers (including NGOs). However, extension workers found there way into service provision through being hired by the NGOs. Some farmers directly demanded the FFS way of teaching from NAADS service providers. In response to farmers' demand, some NAADS coordinators made adjustments in the way the programme was operated in their sub-counties. In Soroti, particularly Abuket sub-county, the change contributed to a modification in Terms of Reference for service providers under NAADS. This was possible because the coordinators at district and sub-county level were once actively involved in FFS under FAO-IPPM and from their experience felt FFS was a more adequate approach to train farmers about improved or recommended technology for improved agricultural production/productivity. As a result of influence from the FFS approach, the previous three months NAADS service provider contract was stepped up to six months so that it covered an entire season, thereby favouring use of FFS methodology. Proposals with FFS as the suggested farmer training model had priority over others in winning NAADS farmer training contracts. Because most service providers were not adequately informed about FFS methodology, some co-opted FFS alumni in their firms. FAO-IPPM farmers benefited most from this. My data do not cover use of FFS elements in NAADS and a follow-up study is needed to further cover this aspect of the wider impact of FFS initiatives. One district coordinator observed that co-option of FFS alumni turned out to be "political". It was used more as a strategy to win contracts because FFS was seen to have prestige with donors. But service providers did not then use the methodology or even the expertise of the FFS alumni.

FFS alumni served as service providers under NAADS. Lack of minimum academic credentials, however, did not allow them bid for service provider contracts in their own right. FFS farmer alumni therefore found themselves sometimes co-opted into firms holding contracts. In Kabale for example, some FFS alumni under the Integrated Disease Management project for potato blight (CIP-IDM) were co-opted to train about potato seed multiplication, potato blight management and effective use of small plots of land. This was so because potato was chosen as one of the priority crops by some groups and tapping practical experience was seen as a better(and cheaper) option than hiring a specialist from either a research or academic institute. FFS farmer alumni were deemed to have a better understanding of the context on the ground. Although sub-contracted FFS alumni contributed to the reputation and performance of such firms, none of them became rich or famous. Their contributions were taken for granted and many times they either received very little pay or payment was not forthcoming.

5.5.3 Promoting the culture of commercialization in farmer groups

FFS encouraged farmers to undertake collective activities (as a group) even after FFS projects had ended. This section explores the collective activities in which some FFS alumni continuously engaged.

Collective marketing of produce
Land was a big problem, and perhaps (in retrospect) an issue FFS ought to have prioritised. Farmers wanting to grow a group crop in one big field for commercial purposes often could not. As a result, each member grew the target crop on their land. In one group (Moruboku FFS) under FAO-IPPM, sorghum and groundnuts were grown by individuals and harvests pooled for better market opportunities. Group labour helped every member with field operations. Post-harvest activities prior to selling were individual responsibilities. In other cases (like Kamusala FFS, also under FAO-IPPM), the mechanism was different. Group member cared for their own crops (millet, groundnuts and cassava), including all labour, and cooperated only when it came to selling. Depending on market price and quantities brought each member was then given a share of the overall income from sales. Nyabyumba united farmers (under CIP-IDM project in Kabale, Western Uganda) also worked like Kamusala but sold its potatoes to a fast foods restaurant called Nandos in the city. Although Nandos offered a lower price (322 Shillings/kg.) than the open market (300-500 Shillings/kg.) the group was assured of a buyer for its produce all year round. Due to limitation in land, production was entirely an individual business, provided quality was ensured. All potatoes were collected at a store where they were sorted and graded. Big potatoes are bagged for Nandos while the small potatoes were sold as seed to the community (at 60,000-140,000 Shillings a 100kg bag, depending on the size of the potatoes (smaller sizes were more expensive).

Provision of quality seed to the farming community

Under the Uganda National Seed Potato Producers Association (UNSPPA), some FFS alumni from CIP-IDM produced and sold improved quality potato seed to farmers. Long distance exacerbated by a hilly landscape, in addition to expense, limited the number of farmers using clean seed to control potato blight. At Kachwekano-Kalengyere research institute, where potato seed was produced, quality seed was too expensive for most potato farmers in Kabale district. Members of UNSPPA bought seed from the research institute annually, multiplied it and sold it on interested farmers at a competitive rate. This not only supplemented the institute's limited supplies, but also saved farmers travelling expenses.

Abuket Sweet Potato Producers and Processors' Association (ASPPA) formed by FFS alumni under CIP-IPPHM multiplied and supplied quality sweet potato vines to farmers within and outside Soroti district. Bigger organizations like World Vision, NAADS and Investment in Developing Export Agriculture (IDEA) made contracts for the vines. Unlike UNSPPA, ASPPA had all its old members in the same village/parish. The same group processed sweet potatoes and sold them to milling companies. Because the chips were not in such demand as expected, the group used some for production of pastries sold as snacks, especially to pupils from neighbouring schools.

Animal rearing as asset accumulation

Some groups, especially under FAO-IPPM and MAK-ISPUH, used group development funds to buy goats used to build "live" asset bases. As explained above, one scheme worked by buying 4-6 goats and sub-dividing them among sub-groups. Every kid was given to a sub-group member until all in sub-groups were covered. However, at some point the cycle was broken. Some members already owning goats did not give the kids to other members. Some groups continued the process over many cycles. Members now have a source of capital, and have even bought or exchanged cows for goats - a source of regular income (from milk) as well as wealth.

However, these multiplier effects remind us that FFS is not in fact self-contained. And it is at times difficult to attribute changes to any one intervention. Farmers, government workers, NGOs, CBOs and local leaders continuously provide agricultural information and some inputs to the same farming communities. Through proactive search and interest in providing service to the community FFS alumni have influenced operation of programs like NAADS, and have become involved in collective action through production and marketing of affordable high-quality planting materials. Such wider system effects are to be welcomed not only because they complement the activities of other organisations, including research and extension, but because they take services closer to the community.

5.6 Concluding remarks

The internal organisation of FFS includes a variety of activities and processes that together play a crucial role in FFS performance. The same applies to the way FFS is embedded in wider structures and activities. The internal organisation looks at how the activities to guide implementation of the curriculum are handled within FFS to promote participation that translates into building farmers' analytical and decision making skill. Crucial elements, presented in this chapter, are AESA, facilitator-farmer relationship and fund management. The way AESA was handled offers evidence that facilitators and farmers did not fully understand the process and objective of AESA. Limiting AESA to collection of quantitative data does not build farmers understanding of the processes or interactions in the ecosystem but makes FFS another form of on-farm research, where researchers manage the experiments and farmers help in collection of specific data. For farmers to understand their eco-system there is need to put as much emphasis on discovering, discussing, analysing and understanding the interactions of the different elements leading to observed outcomes for which data are collected. Facilitators' ability to fit in the communities and enjoy mutual interaction with farmers depends on personal relations and commitment to work with the farmers. Situations that increase social distance between farmers and facilitators decrease the involvement of farmers with the content of the intervention. Good personal relationships, a clear funding structure and mutual appreciation are important in motivating effective relationship between farmers and facilitators in FFS. Although the fund management system aims to build entrepreneurial skills and motivation for the group to work as one, it is open to abuse by individuals. Institution of regular and verified accountability by the group for the funds they received in monetary and "technical" form may minimize such abuses. All these problematic factors have a common source in the management of social capital in FFS. FFS projects depend on well-functioning groups with sufficient levels of mutual trust, commitment and shared interests. Creation of such groups is usually done through investments, attracting free riders, rent-seekers and individuals with out-of-line motives undermining group formation. To circumvent this problem is difficult, but linking existing social formations of a similar size and solidarity might work out better. For this reason some have chosen to try and "embed" extension initiatives such as FFS within other organizational networks (e.g. existing schools or faith-based organizations)

Another issue is the connection between lower and higher levels in the FFS programs. This connection is mainly established through three activities, reporting, field days and field visits or tours. The processes of integration make for greater upward than downward accountability. A technically-oriented one-way reporting system, reflecting the interests of donors and researchers, limits information flow on processes at lower levels that might otherwise be used as input for improving the model. Interactions between farmers, content and facilitator are important in identifying and devising areas for improvement. Using field days to please higher level managers defeats the objective of FFS, focused on farmer learning. Reporting on field days and field visits will only benefit participation when the higher managerial or policy levels take up challenges emerging from the field in a serious way. FFS is not independent of activities and structures in the wider community. No contributions or improvements can be solely attributed to a particular intervention, because every new intervention builds on something developed previously in some way or other. Contributions from farmers, government organisations, NGOs and CBOs in any field complement what FFS offers to the community, perhaps most evidently in the fields of animal husbandry, and marketing as well as the agronomy of crops not promoted under FFS. Influence of other programmes and community activities on proactive

FFS alumni not only move services closer to the people but also complements research and extension, as well as building up a spirit of partnership over innovation. For this reason FFS should be careful not to separate itself from the wider field of institutional transformation aimed at farmers in Uganda. FFS is far from being a panacea, as this chapter has shown, and there is much scope for mutual learning.

CHAPTER SIX

Discussion and conclusions: FFS as a tool of participatory agricultural extension in Uganda

6.1 Introduction

This thesis has focused on increasing the relevance and hence the functioning of participatory innovation-based agricultural extension projects for poverty-reduction. Agricultural technologies are important elements because farmers (as communities of learners) can employ technology to improve rural livelihoods. For that reason, science and technology are recognized as essential components in strategies for promoting sustainable development (UNDP, 2001b, ICS, 2002). Effectively operating innovation systems can only function properly when the system is geared towards solving prevailing problems in the targeted community or society. This requires technology that is relevant and practical.

An assumption underlying this study is that agricultural extension programs will remain important paths for rural and agricultural development and therefore poverty alleviation. This thesis attempts to bring out areas that need improvement and suggests how improvement of FFS might be achieved if it is to have an impact on poverty in rural communities in Uganda. The key point projected in this thesis is the importance of (pro)active involvement of communities in the creation, diffusion and use of technological solutions.

Participation is used as a tool to enhance people-centred development with the ultimate aim to minimise poverty. In theory, agricultural extension through FFS is an important entry point through which participation is reinforced. The pressure for agricultural extension to be responsive to current challenges of agriculture in Uganda has given rise to client-oriented approaches that support a rural development agenda (Rivera and Alex, 2004). As Uganda reforms its agricultural extension system, it is important to analyse both the old system and the proposed new system in order to make the new system more effective and relevant (Rivera and Alex, 2005). Reflecting on the operation of FFS with the objective of making it more effective and relevant has been the objective of the present study. FFS began in Asia (Chapter one) but has now spread widely (Braun *et al.*, 2006). Although the basic formula travelled across regions, FFS has tended to adapt itself (as this thesis has shown) to existing patterns and styles of extension. This leads to a question whether FFS be scaled-up without losing track of its basic objectives.

When employed appropriately, FFS should lead to a match between demand and supply of technology, resulting in desirable, relevant and effective farming systems, and increased agricultural production. The operation of FFS in mixed and diverse farming systems, as found in Uganda, makes its realization a real challenge. The introduction of FFS is not a neutral instrument but is in itself affected by concurrent processes on various levels. First, the decisions, activities, interests and operation of all organisations (i.e. the formal institutional context) affect the implementation process. The functioning of existing organisations strongly affects the operation of FFS, as has been shown. Second, the local culture of the farming community into which (FFS) projects are introduced also influences how farmers relate to interventions. Farmers tend to see 'external' interventions in the perspective of compatibility with what they do and need. Some of these points are now briefly summarised.

6.2 Main findings of the thesis

This thesis has centred on a technography of FFS in Uganda. It has striven to give an analytic description of activities, processes, relationships and interactions between people and technology. Chapter one analysed the concept of participation in improving upon community development through re-orienting agricultural extension and highlighted the framework within which the study was conducted. In relation to the technographic strategy of this thesis, chapter two focussed on the institutional elements/actors at work in FFS and the external/ internal adaptation process through which FFS in Uganda was set up. The overall point put forward was that FFS inadequately translated into a participatory model and did not fit well in the Ugandan (non-IPM) context. The thesis describes how mandates of research and funding institutions at higher levels took priority in shaping FFS. This precedence of research mandates and interests reduced FFS to a technology transfer model repeating some of the top-down mandatory and instructional failings of earlier conventional extension systems. Using FFS as a tool indirectly to force scientist-based or external ideas on farmers translates FFS into a non-participatory transfer model little different from traditional extension models. This has implications for the relevance of technologies promoted through FFS innovation systems in Uganda. That is why a change in structures, functions and mandates of actors in innovation systems (as suggested in chapter two) is important if technology, generated from within or outside is to fit the specific context in which FFS has to operate.

Interaction of institutions was stimulated by the technology promoted. According to the technographic scheme, description and analysis of the various ranges of interventions covered by FFS in Uganda was undertaken in chapter three, and revealed much "off-shelving" of already manufactured technologies from research stations. Off-shelving as a goal favours an instructive extension approach, in which the uptake of the introduced technology is essentially a gamble. In-situ articulation and development of adequate technology through participatory methods is a difficult and time consuming process. This is not an attractive option for scientists with little or no training in participatory methods. They prefer to offer ready-manufactured technologies, but these are often irrelevant to communities not considered while the technologies were being developed. Pushing a misplaced technology leads to non-use. Attention paid to catalysing, building and promoting local innovations might lead to interventions better meeting local realities, it was argued.

Analysis of the response of communities of learners under the five FFS project interventions examined in detail only confirmed the suspicion that most ready-made technologies are a poor fit with farmers' local practices (chapter four). Farmers are keen to learn but do not take up anything they are exposed to simply because they have been told about it, Farmers more easily take up technologies or ideas that are compatible with what they do, know, need, and are used to. In this case, building on existing local experience is very important in integrating something new into day-to-day activities. Experience, interaction and human activity are the basis of learning. Learning in this thesis has been considered a social process, and relevant interventions successfully build upon or connect to that process. Understanding local traditions and practices, often considered as "traditional and inferior" technology, and the rationale behind the use of such technology, is therefore imperative in generating insights on choice of new, more useful, better and improved interventions. Taking into consideration local people's priority needs improves upon the relevance of such technologies.

Besides the institutional "baggage" inherited from the international scene (chapter two) analysis of the local organisation of FFS projects on the ground and their links with other structures/activities (chapter five) brought out a similar pattern of interactions underlying failure to translate FFS into support for local community capacity. Formalisation of the curriculum, AESA and field days limit active engagement of farmers in discovery-based interactive learning processes through which observation and analytical and decision-making skills can be built, as expected in FFS. Through formalisation, structures at higher levels hijack activities. Objectives, outcomes and procedures of operation reek of the old instructive institutional traditions. Making a report more technical the higher it went, without sharing it to participants or feeding back outcomes, promoted only the interests of donors and researchers. This example revealed clearly that activities and processes linking lower and higher structures were used mainly to aid smooth accountability and advancement of interests of the higher level actors. Integration of proactive FFS alumni in local organisations, however, contributed to improved service delivery to the farming community. Rethinking and improvement of the internal organisation of FFS is necessary to strengthen and promote local capacity, if better technological outcomes are to be achieved.

6.3 Changing FFS to support innovation systems

There is a growing attention for innovation systems as the unit of analysis and focus for change processes in international development (Hall and Dijkman, 2006). Technological change is only one element in the system approach. Furthermore, technological innovation only takes place when system synergies kick in. As research modes and intervention practices become more interdisciplinary and participatory, this has an effect on all actors and process within an innovation system. This also implies that investment in local structure is as necessary as investment in the organisation of agricultural science, since both sets of elements are important influences over the functionality of the innovation system as a whole. Looking at the example of the difference in functioning between farmer facilitators and extension staff facilitators (chapter three and five), it became clear that re-alignment of local and "scientific" structures improves the relevance and functionality of innovation systems in a given local community. Creating protected environments for skill training alone may not make scientists competent in adequately implementing participatory innovations. Participatory innovations for poverty reduction come in many forms. Crucial for all these forms is that technology-oriented innovations incorporate an approach to change in which science and technology continuously interact with prevailing problems in the context of application. The transformations make an innovation system adequately answer the current challenges in agricultural development and therefore enhance the relevance of knowledge investments. Action research (Reason and Bradbury, 2001) provides one model for inducing this increased functionality, since science is cultivated as embedded in culture and not boxed off as an ivory tower activity. Traditional structures undermine innovation pathways by reinforcing old ways of doing things (chapter two), so "demolishing" existing organizational structures and creating new ones is one of way of enhancing relevant operational modifications.

Agricultural research needs a properly designed and well organized channel, in other words a well-functioning innovation system, to reach its clientele. There has to be demand for what innovation systems can offer - i.e. farmers need to campaign for realistic (technology) solutions to farming problems on the ground. FFS can play a role not only as a teaching tool but also as a tool for experimentation that increases demand for innovation among the rural poor. This demand can be cultivated through research that engages in open-ended experiments and follows up on farmers' specific concerns. In open experiments the output to farmers is do-it-yourself in sight into technology, and not a set of instructions to be followed. Scientists then keenly follow up the farmers to understand what farmers actually do with technology, how they modify it, or why they abandon it, thus exposing the rationale behind their actions and decisions. Action research addresses the problem of discourse.

A problematic element of participatory development, as Richards (2006) points out, is the amount of reliance placed upon discursive aspects. Much of technological change is non-discursive. People - whether scientists or farmers - have to do it. An innovation system is at the end of the day performance rather than talk. This performance-oriented nature of technology-induced change is something that is overlooked by many proponents of participatory approaches. So far, very little is known about non-discursive participatory methods for technology-induced change. FFS (in theory) stresses learning-by-doing and discovery-based approaches to problem solving, but in practice relies too much (in the Ugandan applications examined above) on getting farmers to sit down to discuss. This bias towards talk and away from performance converts extension practice back into the model it displaced - Training and Visit (sometimes known by sceptics as the Talk-and-Vanish model of agricultural extension).

A practice-oriented approach, such as FFS aspires to be, is worth persisting with, however, since it improve the chances of developing effective technologies *in situ*. In spite of inconvenience due to resource related constraints (chapter three), FFS challenges scientists to prove themselves wrong when developing technology-based innovations for specific areas rather than proving themselves right through already manufactured technologies. Getting things wrong is, in fact, a fruitful first step towards truly understanding the problem at hand, in its context, and thus of getting it right in future attempts. The constructive and rational nature of farmer decisions and actions (Scoones and Thompson, 1994) offers rich learning opportunities to researchers. Learning from what farmers do, however, requires adequate communication skills, particularly in regard to listening careful to what farmers say and in interpreting actions that appear at times to be confounding or flouting text-book understanding. Communication and facilitation are essential tools not only in inducing and managing changes (Kibwika, 2006), but also in analytically grasping local contexts. Such skills in comprehension are required across the board - by scientists, extensionists and community learners, among other actors, because inducing change in the innovation system is a collective and communicative process. Cornwall *et al.*, (1994) emphasizes need for research and extension to understand varied social contexts within which farming activities are embedded, so as to devise more appropriate methodologies and more relevant technologies. It seems clear, from analysis above, that participants in FFS in Uganda need to re-learn FFS methodology, and to reconceptualize their interventions not as technology transfer but as catalytic action within an innovation system shaped by cultural, personal, and political dimensions.

Innovation systems are open ended. Making FFS interventions open is more likely to increase chances of FFS being more response and effective in solving farming problems. Working with farmers as proponents of participatory action research (Selener, 1997), and building on empirical practices deployed by farmers, is an avenue for development of innovations systems closely adjusted to local realities. Dogmatic dependence on laws and rules (of science) may, by contrast, serve only to keep FFS closed and sometimes misleading. And yet all is not doom and gloom. Its procedures are sufficiently transparent to pinpoint inadequacies and wrong turnings, as this thesis has been able to demonstrate. In other words, FFS is (despite risks of closure) a sufficiently open, dynamic system to sustain a vigorous debate about scope for change and improvement. Thus it fits the requirement of Friedman (2001) for action science that aims at "research in practice not research on practice" (p. 160). Promotion of solution oriented thinking and the practice of deliberately modifying a process to answer prevailing and emerging problems requires commitment and some level of motivation. At times this is too much of a challenge for scientists motivated by professional concerns for advancement through publications based on generalization rather than on documenting interventions suited to specific local contexts. This is one of the main challenges in the restructuring of FFS as a potential component within effective pro-poor rural innovations systems. To make client-oriented research practices attractive requires adjustment of quality criteria in the international science system. Unless participatory research is taken fully seriously, researchers will not consider it science that can be published. Realizing the strength of action research requires a change in research procedures. Working and building on farmers' practices motivates scientists in research and extension to identify what works, what does not work and how to make it work. The burden of developing technology fitting farmers' problems does not lie on scientists alone but is a challenge to all actors within a pro-poor innovation system.

Agricultural professionals are crucial actors in innovation systems. As articulated in this thesis, planning and designing (chapter two) and implementation and reporting on FFS (chapter four) was led by agricultural professionals, i.e. researchers and extension workers played the central role. This is why Kibwika (2006) argues that universities need to respond to current complexities and challenges in African agriculture by providing training that produces balanced professionals with an appropriate orientation towards addressing problems in the complex (i.e. inter-disciplinary) form farmers actually encounter them. The balance between science and practice that obtains within the agricultural education system will have a major impact on the practicality and relevance of technology development (Harwood, 2005). An orientation that approaches technology as rule following behaviour, i.e. the correct application of recommended practices as guided by formal scientific principles, allocates farmers to a passive role, and increases the chances of following development paths irrelevant to local contexts or needs. Such orientations inculcate a tendency to look at farmers as a homogenous group, interested in the technology for its own sake, rather than in its use. Technology developed without consideration of the users' perspectives risks wasting effort. To attain sustainable agricultural development there is a specific need to link informal (farmer) and formal (scientific) knowledge to ensure a focus on usage. The two parties bring expertise in different areas, and these complementarities result a richer and more useful resource (Collins and Evans, 2002).

For effective harnessing of different streams of knowledge and expertise within a development-oriented innovation system, Cash *et al.*, (2003) argue that scientific information ought to be credible, salient and legitimate to the users. Inserting a new element like FFS into an innovation system requires careful scrutiny of whether and how well the innovation process is working out for users. Borrowing from Harwood (2005), there is need to reduce the attention paid to basic institutional rules in science (e.g. conventional peer review) and to find more effective ways of assessing (and rewarding) impact in the field. Farmers problems are neither fixed, nor a form of puzzle that can be fixed over night, using available technology as seen in chapter three and four, and it would be interesting to work out what might be needed to create a kind of monitoring and evaluation system sensitive to impacts over a longer period, coupled to the professional reward system for scientists. How this might be done requires a wide debate among scientists and other stakeholder groups, beyond the scope of the present discussion.

But what can be stated with some confidence, summing up the lessons of the technographic account of FFS offered in this thesis is that development of innovation systems adapted to users' contexts will be a continuous and dynamic process, in which action leads to learning and learning feeds further action. Research and learning have to be seen as continuing, dynamic processes, not moments of discovery (Carson and Sumara, 1997; Pedler, 1997). Dynamism within an innovation system results from different ideas coming continuously from the different actors exerting efforts to improve system performance. It is incontestable that science and technology are important potential factors in poverty eradication, but more is involved in agriculture innovation systems that simply making technologies acceptable or useful. FFS must aspire to contribute to these other aspects as well. As Richards (1993) suggests, agriculture is (from a farming perspective) a performance embedded within the wider performance of social life. FFS could well be a crucial kind of intervention, in that it brings farmers and scientists together at an interface where outcomes can be judged by performance standards, and specifically the standard of poverty reduction. But this thesis has shown that specific reformative action is now needed to prevent old, doctrinaire attitudes regrouping under participatory disguise.

6.4 Stepping-stones to an effective FFS-based innovation system

The effectiveness of FFS in re-orienting agricultural extension, this thesis has argued, largely depends on how it is used. Based on the technography of five FFS projects, this thesis suggests some starting points to improve upon the functioning of FFS, to make it more relevant and effective in re-orienting agricultural extension in Uganda, and perhaps else where in Africa. I refer to these lessons as stepping-stones. They are:

1. Understand local context in terms of what people do socially and technically, with reference to targeted crops (technology). Both sets of factors are equally important areas upon which researchers and extension workers need to work, prior to introducing new or improved technology at community level. Many methods can then be applied, but the core requirement in all cases is to work with communities as a means to actively listen to the rationale behind their decisions and activities, thus providing better understanding of current practices and thus arriving at better sense of what needs to be strengthened or changed. The rubric of working with farmers in order to better understand them is also useful as a strategy to mobilize appropriate partners at the community level, and to come to common agreement about best ways of working cooperatively.

2. Resist the temptation to configure FFS around assumptions of technology transfer, and favour instead a process where farmers take an active role in choosing and developing technologies addressing their needs. Developing and choosing technology relevant to specific community realities and needs requires interacting and learning with farmers in question. It also implies building on what farmers know. A technology will be relevant if it is useful in the users' perspective, not the researchers' perspective. As a dynamic and open model for change FFS will frequently undergo modification based on shifts in the prevailing context. Tying it in the conventional and formal settings disables the basic concept of interactive or social learning.

3. Appropriate facilitation requires people-oriented and technical skills. People-oriented skills are not only useful in understanding and analysing existing situations but also in identifying more innovative and appropriate ways of engaging and supporting targeted communities for realisation of their development. Non-discursive approaches - e.g. actual experimentation as opposed to workshops - offer distinct and important learning opportunities to find out what works or does not work.

4. Adequate choice and use of methods is a means effectively to integrate the lower and higher structures for mutual benefit, accountability and firm embedding of FFS in the social world farmers inhabit. Active involvement of all actors in the entire process of diagnosing the problem, designing feasible solutions and implementing them helps to ensure that every actor owns the process and is accountable to it.

5. Change in organizational functioning and mindsets to suit participatory oriented research and extension in agricultural, rural and community development is the overall turning point upon which all the rest depend. Development of appropriate innovation systems of

an FFS kind requires professionals and other actors to develop a positive attitude open to change and to treat farmers as constructive people responsible for their own destiny and development.

To ignore these stepping stones is to risk FFS turning into a kind of science-based cargo cult. In cargo cults, believers imagine that by performing certain formalised steps good things will automatically happen. Melanesian islanders saw European colonialists build ports and jetties to ship their goods. They then imagined that making mock-up equivalent structures would bring similar goods by magic. Simply carrying out an FFS "according to the rules" will have little or no impact on real world processes. It is only if the FFS modality develops true insights into basic processes determining agricultural underperformance and poverty that real changes will occur. The present thesis has shown that while some progress has been made via FFS in Uganda on addressing some agro-technical issues the current weakness of FFS is its inability to analyse and address some of the socio-political mechanisms of class and social differentiation embedded within the practice of science in Africa. It is perhaps only through changing (and not through reproducing) these social inequalities that agro-technical change will achieve its full potential as an instrument of poverty alleviation.

6.5 Concluding remarks

Findings in this study lead to a general conclusion that the way in which FFS was implemented has failed adequately to re-orient agricultural extension systems in Uganda to make them responsive to local problems. The functioning of Ugandan FFS has become caught up in the top-down approach it was meant to improve on (e.g. farmers are treated as receivers rather than active shapers of technology). When participatory approaches like FFS revert to being conventional, they cease to be relevant to poverty alleviation. Perhaps the most important single finding of this thesis is that allowing FFS to become formalised tended to reproduce the social order within the organizations participating in its development. In short, existing organizations change FFS to their own needs, rather than vice versa. But this does not readily show up in assessments. FFS is regarded leniently because it does not threaten the status quo. It is as good a way as any of meeting a demand for accountability and participation without actually reforming the wider institutional landscape. Satisfaction with the status quo does not motivate people and organizations to change. FFS was supposed to challenge organizations (both national and international) to re- examine their cultures, rules, norms, structures and functioning in order to place the clientele for agro-technical innovation in the driving seat. Continuous reflection about the need to develop relevant technology innovations, the role played by farmers and how farmers are supposed to be involved differently in researxh and development, and how to handle public complaints or criticism about methods of operation, among others, is needed if organizations are to begin to recognize the harm caused by the status quo. This recognition does not happen automatically. It calls for leadership and champions (Kibwika, 2006) but also organizational changes capable, intended to trigger and sustain shifts in styles of thought (Douglas, 1986). According to Douglas, it is only when the institution is differently configured that its functionaries will begin to think differently.

But this thesis does not draw wholly negative conclusions about FFS in Uganda. Already, it is clear that it makes a better, more creative and challenging connection between scientists, extensionists and farmers than was achieved under the earlier extension systems it replaced. The present study has offered some evidence that farmers incorporated within FFS have increased their knowledge base about new technologies, and is capable of showing greater cohesion and capacity for collective action. What now needs attention is to ensure that observation-based and practice oriented learning around contexts of collective/community interests or concerns do not fade to something else, specifically to a formalised technology transfer vehicle serving the needs of researchers needing to justify their existence through the take up of innovations emanating from their organizations. It seems clear that in the end what is at stake is an issue of power. The Inter-Academy Council report on science and technology for African agriculture commissioned by the Secretary-General of the UN (Inter-Academy Council 2004) includes a statement made by one of the regional consultation groups (Appendix A) to the effect that local farmer organizations in Africa are weak. This is manifestly true. Perhaps it is only when such organizations are strengthened, and show interest in participatory approaches to technology development, that FFS in Uganda will achieve its full transformative potential.

Annexes

Annex 1: Table showing inventory of FFS projects in Uganda

Organizations using FFS methodology	Period when operational	Technology promoted	Coverage
FAO (Food and Agriculture Organization)	1999-2001 (phase I)	Integrated Production and Pest Management (IPPM) in cotton, vegetables, groundnuts, maize, bananas	Regional program in Eastern Uganda (Soroti and Busia), Western Kenya (Kakamega, Bungoma and Busia)and Northern Tanzania
Second phase (FAO-IPPM)	Sep 2005-2008 (phase 2)	Any crop as desired by FFS groups but with more emphasis on Marketing and networking.	Add Kumi and Bugiri as well as Mozambique
CIP-NARO (CIP-IDM)	1999-2001	Integrated Disease management in potatoes (IDM)	Kabale (Western Uganda)
Makerere University, faculty of Agriculture (MAK-IPM)	2001-2002	Integrated Pest management (IPM) in groundnuts and cowpea	Iganga, Pallisa and Kumi (Eastern Uganda)
(MAK-SPUH)	2004-2005	Safe pesticide use and handling (SPUH)	Mukono, Kiboga (Central Uganda) and Mbarara
(MAK-INM)	2003-2005 June	Integrated nutrient management (INM)	Pallisa (Eastern Uganda)
CIP-NARO (CIP-IPPHM)	2002-2004	Sustainable production and post harvest management of sweet potatoes	Soroti (Eastern Uganda), western Kenya and Northern Tanzania
FAO-Africa 2000network (A2N-ISPI)	2003-2004	Integrated Soil productivity improvement (ISPI)	Tororo and Busia (Eastern Uganda)
Danish Assistance to the Self-Reliance Strategy (DASS) (DASS-IPPM)	2002-2005 June	Integrated Production and Pest management (IPPM) and Sustainable agriculture based on low external input	Adjumani (Northern Uganda)
Environmental Alert (Envalert-INM)	2003-2005	Integrated Nutrient management (INM)	Wakiso (Central Uganda)
FAO-NARO (FAO-CA)	2003-2004	Conservation Agriculture (CA)	Mbale and Pallisa

Annex 2: Criteria for assessing project/programme compliance with PMA.

Tick the PMA Priority Area (s) addressed by the projects/programmes

PMA Priority Area	Project No. (Tick whichever applies).						
	1	2	3	4	5	6	7
1. Research and technology development							
2. Provision of advisory services							
3. Agricultural education							
4. Access to rural financial services							
5. Natural resources use and management							
6. Agro-processing and marketing							
7. Supportive physical infrastructure							
8. Other (specify)							

No Review criteria	Score								
	- -	1	2	3	4	5	6	7	

I. Extent to which project contributes to the aims and objectives of PMA (70%).

1. *Increase in Incomes and Improvement of Quality of life of the Poor*	30								
a. Is the project targeting the needs and constraints of the vulnerable groups such as subsistence farmers, women, widows, female headed households, people with disabilities, youth, orphans, elderly and the sick (HIV/AIDS)?	8								
b. Will the Project build capacity of the poor/community to benefit directly from and take up new opportunities offered by the Project to improve their well being?	5								
c. Is the project gender sensitive and responsive to the constraints and needs of the community (offering equal opportunities for both men and women)?	5								
d. Are project resources/funds allocated to poverty eradication activities/interventions?	4								
e. Does the project address a strategic intervention, opportunity or challenge, such as reduction of losses and value addition?	4								
f. Does the project empower the target group to continue activities after the project ends?	4								

2. **Household Food and Nutrition Security Directly and/ Or Through the Market**	10								
a. Will the project build household capacity for increased productivity?	2								
b. Does the Project encourage production for the market?	3								
c. Is household ability to purchase food from the market enhanced?	5								
3. **Provision of gainful employment**	15								
a. Does the Project promote job creation?	5								
b. Does the Project promote income generation?	5								
c. Are there opportunities for secondary benefits such as agro-processing and improvement of market infrastructure?	5								
4. **Sustainable Use and Management of Natural Resources?**	15								
a. Project will conserve natural resources such as soil, water and forests.	5								
b. Project will not negatively affect the environment	5								
c. Project will promote sustainable use and management of natural resources.	5								

II. Implementation Within the PMA Policy Framework and Principles (30%).

1. **Provision for Multi-stakeholder participation in Initiation, Design, Implementation, Monitoring and Evaluation.**	15								
a. Participation of the target group (s)	4								
b. Participation of the local governments, lower councils and communities.	3								
c. Participation of the civil society	3								
d. Participation of the private sector	3								
e. Adequate linkages with other stakeholders and service providers such as training institutions and farmer groups	2								
2. **Can the Project/Programme be Funded under the PMA Arrangements (e.g. Government Grants to Sub Counties, Ministries and Agencies)?**	15								
a. Project focuses on public goods and/or services such as multiplication of foundation seed, demonstration plot showing improved technology and repair of a bridge on community road.	8								
b. Resources are allocated to activities that ensure timely and quality delivery of outputs.	2								
c. Ways and means for proper use and accountability of project funds are clear and adequate	3								

d. Ways and means for monitoring, evaluation and reporting are clearly outlined.	2									

GRAND TOTAL SCORED	**100**

Decision: (1) Project Scoring 60% and above: - PMA compliant/compatible if there are no issues requiring project/programme realignment. Action: Approve the project/programme, (2) Project scoring 60% and above: PMA compliance conditional if there are issues that require project/programme redesign, refocus and/or adjustment in order to be realigned. Action: Differ approval of the project/programme until project is fully realigned. (c) Project scoring below 60%: - Not PMA compliant/compatible. Action: Realign the project/programme.

Annex 3: List of FFS groups visited, their sub-counties and districts

District	Name of FFS	Village	Subcounty
Mukono	Kasayi FFS	Kasayi	Kyampisi
	Balikyewunya Development Association FFS	Kirowoza	Goma
Kiboga	Kakunyu C FFS	Kakunyu C	Bukomero
	Kyosimba Onanya Women's group	Kawaawa	Lwamata
	Kiddawalime farmers' Pesticide FFS	Mulagi	Mulagi
	Kazinga FFS	Kizinga and Kajeera	Kibiga
Mbarara	Nyakarambi-Bugonzi FFS	Nyakarambi	Nyakitunde
Iganga	Kiboyo FFS	Kiboyo	Nakigo
	Mwinho Akuwa Tweyambe group	Buwolomera	Bulamagi
Tororo	Molo FFS	Molo	Kisoko
	Karwok FFS	Karwok	Molo
Busia	Sihubira FFS	Sihubira	Lunyo
	Bulime FFS	Bulime	Lunyo
	Ndegero FFS	Ndegero	Lunyo
	Busiime FFS		Lunyo
	Jirani FFS	Mundaya	Busitema
	Kawabona kabosi FFS		Lunyo
	Buyaya FFS	Buyaya	Lunyo
	Mwendapole FFS		Dabani
	Mundaya FFS	Mundaya	Dabani
	Huliime FFS		Lunyo
	Bruda FFS		Lunyo
	Lwala FFS		Lunyo
	Buwumba Elders Association		Dabani
	Malanga FFS	Busime	Lunyo
	Buwumba Elders' Development Association		Dabani
	Sihulawula		Dabani
	Mugungu FFS		Town Council
	Sikada Elders' Development Association		Dabani
	Busabale FFS		Dabani
	Bakisa FFS		Lunyo
	Njala Siyitya FFS		Masafu
Kumi	Akuoro United Farmers	Bukedea	Bukedea
	Olupe FFS	Olupe village	Olupe
Soroti	Abuket FFS	Abuket	Kyere
	Okunguro FFS	Okunguro A	Kyere

Asianut Womens' group	Angole	Kyere
Apamora FFS	Angole	Kyere
Ogobbai FFS	Ogobai	Arapai
Obak FFS		Kamuda
Madera United Farmers	Madera	Madera
Omodoi FFS	Omodoi	Katine
Akisim FFS	Kyere	Kyere
Asuret Womens' group	Asuret	Asuret
Omolo FFS	Omolo	Katine
Abaringentie FFS		Katine
Moruboku FFS		Bugondo
Kamusala FFS		Kateta
Tong Piny FFS		Arapai
Atamaisi FFS		Kyere

Annex 4: Comparison of FFS and other projects as perceived by farmers

Annex 4 (a): Comparing MAK-SPUH FFS and other initiatives as information sources

Parameter	BUCADEF	UNFA	FFS
Contribution to group formation and functioning	Initiated and motivated working in groups	Worked with very few farmers in groups	Uplifted people to work in groups and exposed others to work in groups
Input provision	Provides only knowledge during training	Provides knowledge and training materials to read. Subsidizes improved seed	Provides knowledge and inputs for learning
Frequency of meeting with the groups	Meeting monthly (Tuesdays) to plan or solicit for ideas at village level. Theoretical	Meeting at parish level not regularly. Often twice a season or less. More theoretical	Meet regularly (at least once a week). Learning is based on hands on experience
Scope	Deals with many crops	Teach many things: crop, animal, and managerial related aspects	Deals with only one crop and emphasizes pest management
Training site	No specific place for training	No specific site for learning	Has a specific site for the learning, and what to talk about

BUCADEF = Buganda Cultural Development Foundation
UNFA= Uganda National Farmers Association

Annex 4 (b): Comparison between FAO-IPPM FFS and Sihubira (a CBO) in Busia

Parameter	Sihubira Farmers' Organization - SFO	FFS (FAO-IPPM)
Use of input from the community	Uses peoples' initiatives and is internally driven. Problem identification is left to the people. Farmers contributed resources: stationery, food/snacks. No money	It is top bottom and is externally driven. People at the top decided on what the problem was and did not engage the community. Program provided stationery, money and snacks
Emphasizes method of training	Used demonstration plots and emphasizes comparison of old and new/improved varieties	Used experimental plots and put more emphasis on Agro Eco-System Analysis (AESA)
Focus of training	Trained about agronomy of different crops (excluding cotton) and mainly food crops	Trained agronomy with more emphasis on pest control in income generating or cash crops
Payment of facilitators	Facilitators not paid - more voluntary	Facilitators paid by the program through farmers
Contribution of group for membership	Encouraged group formation and registration with the Association is with some membership fee	Use already existing groups and no membership fee for the groups to be part of the FFS program
Duration with a group	Supports the group throughout and the group contribute to training of other members in the community	After one year of FFS training, groups are left on their own. However some get taken on by incoming projects whose link is attributed to FFS

Annex 4 (c): Comparison between SOCADIDO, FAO-IPPM FFS and NAADS

Parameter used	SOCADIDO	FFS	NAADS
Input provision	Provides inputs like hoes, improved seed, cows, goats to individual farmers	Provides inputs as learning materials for a group	Provides inputs mainly for demonstration and multiplication purposes
Interactive teaching method	Demonstration sites	Experimental plots, Focus on AESA	Demonstration sites
Frequency of holding (Farmer) meetings	Meet when necessary, especially at beginning of the season	Weekly - regular	Meet once or twice a month when need be
Ease of access with facilitator or service provider	Difficult to get in touch with the extension worker/facilitator	Very easy to get in touch with the facilitator. They keep checking on the farmers very frequently	Difficult to get in touch with the service provider because they keep changing based on the enterprise and are very mobile.
Scope/coverage	Extensive and broad covering a variety of livestock, crop and inputs. What to teach the farmers about or what to give them is mainly decided by the program	Intensive and limited with emphasis mainly on pest management in specific crops. General agronomy is handled. Crop is largely pre-determined by the program	Extensive and broad covering livestock (poultry, goats, bee keeping....), crop, management. What to teach the farmers is what the different interest groups choose
Support to old groups	Continues supporting the old groups especially with inputs. No set period for which the program works with the group	Rarely continues supporting the old groups but can recommend some groups to work with upcoming NGOs or programs.	Continues with the groups. Farmer enterprises and interests keep changing and the program seeks to cater for that. No limit as yet as to how long it will deal with a given group before leaving it to move on its own
Handling of funds	Farmers do not handle money	Farmers are given chance to handle money and pay for the facilitation services	Farmers do not handle any money. They only have to attend the training
Composition of men and women in the groups	Focuses mainly on women and no limit on group size	Encourages some gender mix between men and women in groups and limits group size to 25-30 members	Does not mind the gender composition and size. However, farmers' fora at district level should have 30% women representation. No limit to group size

References

Aben, C., Ameu M., Fris-Hansen and Okoth J. R., 2002. Evolution of Extension Approaches in Soroti District. *Paper prepared and presented for a workshop on Extension and Rural Development: A convergence of views on International Approaches held 12-15 November 2002,* Washington D.C.

Abidin, P.E., van Eeuwijk, F.A., Stam, P., Struik, P.C., Malosetti, M., Mwanga, R.O. M., Odongo, B., Hermann, M. and Carey, E.E. (2005) Adaptation and Stability Analysis of Sweet potato Varieties for Low-input Systems in Uganda. *Plant Breeding* 124 (5): 491-497.

Adipala, E., Nampala, P., Karungi J and Isubikalu P (2001) Options in the Management of Cowpea Pests: Experiences from Uganda. *International Pest Management Reviews* 5: 185-196.

Africare (2003) Annual report pp 40 www.africare.org/about/annualreport/2003/2003annualreport.pdf

Agemo, J. (1980) *Social Aspects Associated with Culture Among the Iteso.* Bachelor of Arts Dissertation submitted to Makerere Univeristy, Kampala, Uganda.

Akwang, A., Okalebo, S. and Oryokot, J. (1998) *Needs assessment for Agricultural Research in the Teso Farming System 'Aijul Eode'* main report, NARO-SAARI

Altieri, M.A. (1995) *Agroecology: The Science of Sustainable Agriculture.* West View Press, Boulder.

Anderson, R.J. and Feder, G. (2004) Agricultural Extension: Good Intentions and Hard Realities. *The World Bank Research Observer* 19 (1): 41-60.

Aniku, J.R.F. (2001) Soils Classification and Pedology. In: J. K. Mukiibi (ed.) *Agriculture in Uganda, Vol. 1. General Information.* Fountain Publishers: Kampala, Uganda. pp 66-103.

Anonymous (2003) Promotion of Sustainable Sweet Potato Production and Post Harvest Management through Farmer Field School in East Africa. *Report of an Evaluation/Planning Workshop - II (Report 2) held in Blue York Hotel, Busia, Kenya 1-3 April 2003.*

Arwa, K. (2002) *Assessing the Long-Term Impact of FFS Training on Farmers Knowledge, Attitude, Practice and Empowerment: A case Study in Gezira Scheme, Central Sudan.* M.Sc. Thesis Wageningen Agricultural University, The Netherlands.

AT Uganda (2004) *The Input Distribution Sector at a Glance: summary results of the 2004 National agro-input dealers census.* Report Prepared by Appropriate Technology (AT) Uganda Ltd.

Atingi-Ego, M. (2005) Budget Support, Aid dependency, and Dutch Disease: The case of Uganda. Paper Presented during a *Practitioners' Forum on Budget Support, Cape* Town, South Africa, May 5-6 2005.

Atinyang, K. (1975) *Life in Teso,* MA Thesis, Makerere Univerity.

Babbie, E. (2001) *The Practice of Social Research (9th ed.).* Wardsworth, Belmont.

Bashasha, B., Mwanga, R., Ociti P'Obwoya, C. and Ewell, T.P. (1995) *Sweet Potato in the Farming and Food Systems of Uganda: A farm Survey Report, Sub Saharan Region.* NARO Kampala, Uganda.

Bashasha, B., Nalukenge, I. and Lacker, C. (2004) *Documentation the Contributing Contribution of Contribution of the National Agricultural Research Organization (NARO) to the national Agricultural Policy Process in Uganda.* Report prepared for NARO/DFID/COARD, SAARI by Department of Agricultural Economics and Agribusiness, Makerere University pp 61.

References

Bekunda M.A., Bationo, A. and Ssali, H. (1997) Soil fertility Management in Africa: A Review of Selected Research Trials. In: Buresh R. J, Sanchez P. A. and Calhoun F (eds.), *Replenishing Soil Fertility in Africa*. Soil Science Society of America and ICRAF, Special publication 51, Madison, pp 63-79.

Belay, K. and Abwbew, D. (2004) Challenges Facing Agricultural Extension Agents: A Case Study from South-Western Ethiopia. *African Development Review* 16 (1): 139-168.

Bellah, N.R. (2005) Durkheim and Ritual. In: Jeffrey C. Alexander and Philip Smith (Eds.), *The Cambridge Companion to Durkheim*. Cambridge University Press, UK pp 183-210.

Bernard, R.H. (1995) *Research Methods in Anthropology: Qualitative and Quantitative Approaches (2nd ed)*. AltaMira Press, England.

Booth, D. (2003) Introduction and Overview: Are PRSPS Making a Difference? The African Experience. *Development Policy Review* 21 (2): 131-159.

Braakman, L. and Edwards, M. (2002) *The Art of Building Facilitation Capacities: A Training Manual*. RECOFTC, Bangkok.

Braun, A., Jiggings, J., Röling, N., van den Berg, H. and Snijders, P. (2006) *A Global Survey and Review of Farmer Field School Experiences*. Report Prepared for the International Livestock Research Institute (ILRI), September 2006 pp 91.

Bruin, C.A.G. and Meerman, F. (2001) *New ways of developing Agricultural Technologies: The Zanzibar Experience with Participatory Integrated Pest Management*. Wageningen University and Research Center/CTA, Wageningen.

Bryceson, D.F. (Ed.) (2002) *Alcohol in Africa: Mixing Business, Pleasure and Politics*. Heinemann, Portsmouth, NH.

Bunyatta, D.K., Mureithi, J.G., Onyango, C.A. and Ngesa, F.U. (2005) Farmer Field School as an Effective Methodology for Disseminating Agricultural technologies among Small-scale Farmers in Trans-Nzonia District, Kenya. *Proceedings of the 21st Annual Conference of the Association for International Agricultural Extension and Education*, San Antonio, TX, 515-526.

Bukenya, C. (2007) Stakeholder Participation in the National Agricultural Advisory Services in Uganda: analysis of practice and contextual factors. P.hD. thesis, Wageningen University (*forthcoming*).

Byerlee, D. (2000) Targeting poverty Alleviation in Priority setting for Agricultural Research; *Food Policy* 25(4): 425-445.

Calvert, G.M., Plate, D.K., Das, R., Rosales, R., Shafey, O., Thomsen, C., Male, D., Beckman, J., Arvizu, E. and Lackovic, M. (2004) Acute Occupational Pesticide-Related Illness in the US, 1998-1999: Surveillance findings from the SENSOR-Pesticides Program. *American Journal of Industrial Medicine* 45:14-23.

Carson, T.R. and Sumara, D. J. (1997) *Action Research as a Living Practice*. Peter Lang, New York.

Cash, W.D., Clark, C.W., Alcock, F., Dickson, M.N., Eckley, N., Guston H.D., Jager, J. and Mitchell B.R. (2003) Knowledge systems for sustainable development. *PNAS* 100 (14): 8086-8091.

Chambers, R. (1997) Whose Reality counts? Putting the First Last. Intermediate Technology Publications, London.

Chambers, R. (2002) *Participatory Workshops: A Sourcebook of 21 sets of Ideas & activities*. Earthscan Publications Ltd, London.

Cherp, A., George, C. and Kirkpatrick, C. (2004) A Methodology for Assessing National Sustainable Development Strategies. *Environment and Planning C: Government and Policy* 22: 913-926.

Cheru, F. (2006) Building and supporting PRSPS in Africa: What has Worked Well so Far? What needs Changing? *Third World Quarterly* 27 (2): 355-376.

Christian Aid (2001*) Ignoring the Experts: Poor People's Exclusion from Poverty Reduction Strategies.* Christian Aid Policy Briefing.

CIETinternational (1996) *Baseline Service Delivery Survey: In support of Results Oriented Management in the Uganda Institutional Capacity Building Project.* Final report, February 1996.

Cleaver, F. (2002) Institutions, Agency and the Limitations of Participatory Approaches to Development. In: Cooke, B. and Kothari, U. (eds.) *Participation: The New Tyranny?* Zed Books Ltd, London pp36-45.

Collins, M.H. and Evans, R. (2002) The Third Wave of Science Studies: studies of expertise and experience. *Social Studies of Science* 32 (2): 235-296.

Collinson, M (Ed.) (2000) *A history of Farming Systems Research.* CABI International, Wallingford pp 432.

Conley, T. and Udry, C. (2001) Social Learning Through Networks: the adoption of new agricultural technologies in Ghana. *Amer. J. Agr. Econ* 83 (3): 668-673.

Cooke, B. and Kothari, U. (eds.) (2001) *Participation: The New Tyranny*? Zed Books Ltd, London.

Cornwall, A. (2004) Spaces for Transformation? Reflections on Issues of Power and Differences in Participation in Development. In: Hickey, S and Mohan, G. (eds.) *Participation: From Tyranny to Transformation? Exploring New Approaches to Participation in Development.* Zed Books Ltd, New York.

Cornwall, A. and Brock, K. (2005) What do Buzz Words do for Development Policy? A critical look at 'Participation', 'empowerment' and 'Poverty reduction' *Third World Quarterly* 26 (7): 1043-1060.

Cornwall, A., Guijt, I. and Welbourn, A. (1994) Extending the Horizons of Agricultural Research and Extension: Methodological Challenges. *Agriculture and Human Values* 11 (2-3): 38-57.

Council for Agricultural Science and Technology 1982 *Integrated Pest Management Report No. 93* 105pp.

Creswell, J.W. (2003) Research Design: Qualitative, Quantitative and Mixed Methods Approaches (Second Edition). Sage Publications, Thousand Oaks.

CRF (2001) Farmer Participatory Research on Integrated Crop Management for Sweet Potato in Northeastern Uganda. *Competitive Research Facility (CRF) Project R7024(C) Report.* Department for International Development (DFID), Competitive Research Facility Project R7024[C]. UK.

Davis, K. (2006) Farmer field Schools: A Boon or Bust for Extension in Africa? *Journal of International Agricultural Extension and Education* 13 (1): 91-97.

Deininger, K. and Okidi, J. (2003) Growth and Poverty Reduction in Uganda 1999-2000: Panel data Evidence. *Development policy Review* 21 (4): 481-509.

Delve, R., Miiro. R., Kabuye, F., Kigali, D., Mukalla and Mbaziira, J.S. (2003) *Farming Systems and Soil Management in Tororo District: Baseline Survey conducted and prepared for The Integrated Soil Productivity Initiative through Research and Education (INSPIRE),* July 2003 pp42.

Dewalt, M.K., Dewalt, R.K. and Wayland, B.C. (1998) Participant Observation. In: Bernard H Russell (Ed.) *Handbook of Methods in Cultural Anthropology.* Altamira Press, Oxford pp259-299.

References

Dinham, B. (2003) Growing Vegetables in Developing Countries for Local Urban Populations and Export Markets: problems confronting small-scale producers. *Pest Manag Sci* 59:575-582.

Douglas, M. (1986) *How Institutions Think*. Routledge & Kegan Paul, London.

Douglas, M. (1996) *Thought styles: Critical Essays on Good Taste*. Sage Publications, London.

Ebregt, E., Struik, P.C., Abidin, P.E. and Odongo, B. (2004) Farmers' Information on Sweet potato Production and Millipede Infestation in North-eastern Uganda. I. Associations between Spatial and Temporal Crop Diversity and the Level of Pest Infestation. *NJAS* 52 (1): 47-84.

Ebregt, E., Struik, P.C., Odongo, B. and Abidin, P.E. (2005) Pest Damage in Sweet Potato, Groundnut and Maize in North-eastern Uganda with Special Reference to Damage by Millipedes (Diplopoda). *NJAS* 53 (1): 49-69.

Ecobichon, J.D. (2001) Pesticide Use in Developing Countries. *Toxicology* 160: 27 - 33

Eddleston, M., Karalliedde, L., Buckley, N., Fernando, R., Hutchinson, G., Isbister, G., Konradsen, F., Murray, D., Piola C.J., Senanayake, R., Sheriff, R., Singh, S., Siwach, B.S. and Smit, L (2002) Pesticide Poisoning in the Developing World - a minimum pesticide list. *Lancet* 360: 1163-1167.

Emudong, P.P.C. (1974) *The Iteso: A Segmentary Society under Colonial Administration 1897-1927*. Master of Arts Thesis submitted to Makerere Univerisity. Kampala, Uganda

Emeetai-Areke, T.E.E, Omadi, R., Okwang, A.D. and Eryenyu, A. (2004) *Participatory Breeding/ Selection of Pigeon peas and Cowpeas for Yield and Pests Resistance. Final Technical Project Report, Oct, 2004*. NARO-SAARI, Uganda.

Fakih, M., Rahardjo, T., Pimbert, M., Sutoko, A., Wulandari, D. and Prasetyo, T. (2003) *Community Integrated Pest Management in Indonesia: Institutionalising Participation and People Centred Approaches*. International Institute for Environment and Development and Institute for Development Studies.

FAO (1986) *International Code of Conduct on the Distribution and Use of Pesticide*. Food and Agriculture Organization of the United Nations, Rome.

FAO (1967) *Report of the First Session of the FAO Panel of Experts on Integrated Pest Control* 18-22 September 1967. Italy, Rome Pp 19.

Feder, G., Muragi, R. and Quizon, B.J. (2004) The Acquisition and Diffusion of Knowledge: The case of Pest Management Training in farmer Field Schools, Indonesia. *Journal of Agricultural Economics* 55 (2): 221-243.

Forgas, P.J. and Williams, D.K. (2001) Social Influence: Introduction and Overview. In: Forgas P.J. and Williams, D.K. (eds.) *Social influence: Direct and Indirect processes*. Psychology Press, Philadephia PA pp 3-24.

Friedman, J.V. (2001) Action Science: Creating Comunities of Inquiry in Communities of Practice. In: Reason, P. and Bradbury, H (eds.), Handbook of Action Research: Participative Inquiry and Practice. Sage Publications, London pp 159-170.

Goldstein, H. 1981. *Social Learning and Change: A Cognitive Approach to Human Services*. Tavistock Publications, Newyork.

Gottschalk, R. (2005) The Macro Content of PRSPs: Assessing the Need for a More Flexible Macroeconomic Policy Framework. *Development Policy Review* 23 (4): 419-442.

Groeneweg, K. and Tafur, C. (2003) Evaluation in FFS: A Burden or a Blessing? In: *CIP-UPWARD Farmer Field Schools: Emerging Issues and Challenges. International Potato Center- User Perspectives with Agricultural Research and Development*, Los Baños, Lagun, Philippines pp 298-307.

Gurr, G.M., Wratten, S.D. and Luna, J.M. (2003) Multi-functional Agricultural Biodiversity: Pest Management and other Benefits. *Basic and Applied Ecology* 4 (2): 107-116.

Hall, A. (n.d.) The Origins and Implications of Using Innovation Systems Perspectives in the Design and Implementation of Agricultural Research Projects: Some Personal Observations. www.innovationstudies.org/docs/Andean%20Region%20Workshop%20Paper.pdf

Hall, A. and Dijkman, J, (2006) Capacity Development for Agricultural Biotechnology in Developing Countries: Concepts, Contexts, Case studies and Operational Challenges of a Systems Perspective. *Working Paper Series 2006-003.*

Harwood, J. (2005) *Technology's Dilemma: Agricultural Colleges between Science and Practice in Germany, 1860-1934.* European Academic Publishers, Bern.

Hazell, P. and von Braun, J. (2006) Aid to Agriculture, Growth and Poverty Reduction. *EuroChoices* 5(1): 6-13.

Henriques, P. 2002. Peace without Reconciliation: War, Peace and Experience among the Iteso of East Uganda. Ph.D. Thesis. University of Copenhagen.

Hergenhahn R.B. (1988) *An Introduction to Theories of Learning.* Prentice-Hall International, New Jersey.

Hickey, S. and Mohan, G. (eds.) (2004) Participation: *From Tyranny to transformation? Exploring new Approaches to participation in Development.* Zed Books Ltd, London.

Hirsch, E.D. 2002. Classroom Research and Cargo Cults. *Policy Review*, No. 115. www.hoover.org/publications/policyreview/3459336.html

Humphreys, M., Masters, A.W., and Sandbu, M.E. 2006. The Role of Leaders in Democratic Deliberations: Results from a Field Experiment in São Tomé and Príncipe. *World Politics* 58: 583-622.

INMASP (2002) Integrated Nutrient Management to attain Sustainable Productivity Increases in East African Farming Systems. *Proceedings of INMASP inception workshop 11-15th February, 2002.* Wageningen, the Netherlands pp 37.

Inter Academy Council (2004) *Realizing the promise and Potential of African Agriculture: Science and Technology Strategies for Improving Agricultural Productivity and Food Security in Africa.* IAC Report, June 2004.

ICS (2002) *Science and Technology for Sustainable Development*, International Council for Science (ICS), Paris.

Isaken, A. (2001) Building Regional Innovation Systems: Is Endogenous Industrial Development Possible in the Global Economy? *Canadian Journal of Regional Science* 24 (1): 101-120.

Isubikalu, P. (1998) Understanding Farmer Knowledge of Cowpea Production and Pest Management: A Case Study of Eastern Uganda. M.Sc Thesis, Makerere University. Kampala Uganda 158p

Isubikalu, P., Erbaugh, J.M., Semana, A.R. and Adipala, E. (1999) Influence of Farmer Production Goals on Cowpea Pest Management in Eastern Uganda: implications for developing IPM programmes. *African Crop Science Journal* 7(4): 539-548.

References

Janvry, A.D. and Kassam, A.H. (2004) Towards a Regional Approach to Research for the CGIAR and its Partners. *Expl Agric.* 40: 159-178.

Jehn, K. (1997) A Qualitative Analysis of Conflict Types and Dimensions in Organizational Groups. *Administrative Science Quarterly* 42: 530-557.

Jehn, K. (1995) A Multimethod Examination of the Benefits and Detriments of Intra-group Conflict. *Administrative Science Quarterly* 40: 256-282.

Jehn, K.A. and Mannix, E. (2001) The Dynamic Nature of Conflict: A longitudinal study of Intra-group Conflict and Group Performance. *www.aom.pace.edu/amj/April 2001/jehn.pdf*

Jeyaratnam, J. (1990) Acute Pesticide Poisoning: A Major Health Problem. *World Health Statistics Quarterly* 43 (3): 139-144.

Jusu, M.S. (1999) *Management of genetic variability in rice (Oryza sativa L. and glaberrima Steud) by farmers in Sierra Leone.* Ph.D. thesis, Wageningen University, the Netherlands.

Kai-yun, X and Yi, W. (2001) CIP Potato Late Blight Research in China. *Journal of Agricultural University of Hebei, April 2001.*

Kappel, R., Lay, J. and Steiner, S. (2005) Uganda: No More Pro-poor Growth? *Development Policy Review* 23 (1): 27-53.

Karp, I. (1978) *Fields of Change among the Itesot of Kenya.* Routledge, London.

Karungi, J., Adipala, E., Kyamanywa, S., Ogenga-Latigi, M.W., Oyobo, N. and Jackai, L.E.N (2000) Pest Management in Cowpea. Part 2. Integrating Planting Time, Plant Density and Insecticide Application for Management of Cowpea Field Pests in Eastern Uganda. *Crop Protection* 19 (4): 237 - 245.

Karungi-Tumutegyereize, J. and Adipala, E. (2004) *Enhancing the Role of Makerere University in Technology development and dissemination.* A report submitted to i@mak.com Makerere University pp 51.

Kasente, D., Lockwood, M., Vivian, J. and Whitehead, A. (2002) Gender and the Expansion of Non-traditional Agricultural Exports in Uganda. In: Razavi, S., (ed.) *Shifting Burdens: Gender and Agrarian Change under Neoliberalism.* Kumarian Press, UNRISD.

Kayanja, V. (2003) *Private Serviced Extension goes to the 'Botanical Garden': Multiple Realities of Institution Building and Farmers' Need Selection During the Transition Process in Mukono District, Uganda.* M. Sc. Thesis Wageningen University, Netherlands.

Keeble, D. and Wilkinson, F. (1999) Collective Learning and Knowledge Development in the Evolution of Regional Clusters of high Technology SME in Europe. *Regional Studies* 33: 295-303.

Kenmore, P.E. (1991) *Indonesia's Integrated Pest Management - A Model for Asia: How Rice Farmers Clean up the Environment, Conserve Biodiversity, Raise More Food, and Make Higher Profits.* FAO, Rome pp.56.

Kenmore, P.E. (1996) Integrated Pest Management in Rice. In: Persley, G. J. (ed.) Biotechnology and Integrated Pest management; CAB International, Wallington pp 76-97.

Kibwika, P. (2006) *Learning to Make Change: Developing Innovation Competence in Recreating an African University for the 21st Century.* Ph.D. thesis, Wageningen Agricultural University, Netherlands.

Kidd, A.D. (2004) Extension, Poverty and Vulnerability in Uganda. In: Christopholos, I and Farrington, J. (eds.) *Poverty, Vulnerability and Agricultural Extension: Policy Reforms in a Globalizing World.* Oxford University Press, New Delhi pp 124-170.

Knowles, S.M & Associates (1984) *Andragogy in Action: Applying Modern Principles of Adult Learning.* Jossey-Bass, San Francisco, Carlifornia.

Kogan M. (1998) Integrated Pest Management: Historical Perspectives and Contemporary Development. *Annu. Rev. Entomol.* 43: 243-270.

Kolb, D.A. (1984) *Experiential Learning: Experiences as the Source of Learning and Development.* Prentice Hall, New Jersey.

Lave, J. (1995) Teaching, as Learning, in Practice. *Paper presented as the Sylvia Scriber Award Lecture of Division C: Learning and Instruction at the American Educational Research Association Annual Meeting, April 1995,* San Francisco.

Lave, J. and Wenger, E. (1991) *Situated Learning: Legitimate Peripheral Participation.* Cambridge University Press, New York.

Law, J. (1992) Notes on the Theory of the Actor-Network: Ordering, Strategy, and Heterogeneity. *Systems Prcatice* 5 (4): 379-393

Lawson, D. (2003) *Gender Analysis of the Uganda National Household surveys 1992-2003*, October 2003. Kampala, Uganda

Leeuwis, C. and van den Ban, A. (2004) *Communication for Rural Innovation: Rethinking Agricultural Extension.* Blackwell Science, Oxford.

Leeuwis, C. (2000) Reconceptualizing Participation for Sustainable Rural Development: Towards a Negotiation Approach. *Development and Change* 31: 931-959.

Leeuwis, C. and Pyburn, R. (eds.) (2002) *Wheelbarrows full of frogs: Social Learning in Rural Resource Management.* Koninklijke Van Gorcum, The Netherlands.

Lipsky, M. (1980) Street-level Beauracracy: Dilemmas of the Individual in Public Services. Rusell Sage Foundation, New York.

Long, N. and Long, A. (eds.) (1992) *Battle Fields of Knowledge: The Interlocking of Theory and Practice in Social Research and Development.* Routledge, London.

Low, W.J. (1996) *Prospects for sustaining Potato and Sweet Potato Cropping Systems in SouthWest Uganda.* www.cipotato.org/market/pgmrprts/pr95-96/program

Lynam, J. (2002) Book Review: A history of Farming Systems Research. Ed. Collinson, M. CABI Wallingford. *Agricultural Systems* 73: 227-232.

MAAIF (1995) *Agricultural Extension Project Mid Term Review Report* May 1995. Ministry of Agriculture, Animal, Industry and Fisheries (MAAIF), Entebbe-Uganda.

MAAIF and MFPED (2003) *Guidelines for project/programme submission for PMA compliance and clearance for funding.* Ministry of Agriculture, Animal Industry and Fisheries (MAAIF) and Ministry of Finance Planning and Economic Development (MFPED), Kampala Uganda.

Mancini, F. (2006) *Impact of Integrated Pest management Farmer Field Schools on Health, Farming Systems, the Environment, and Livelihoods of Cotton Growers in Southern India.* Ph.D. Thesis, Wageningen Agricultural Uinversity. Netherlands.

Matteson, P.C. (2000) Insect Pest Management in Tropical Asian Integrated Rice. *Annu. Rev. Entomol.* 45: 549-574.

References

McGee, R. (2002) *Assessing Participation in Poverty Reduction Strategy Papers: A Desk-based Synthesis of Experiences from Sub- Saharan Africa.* IDS Research Report 52. Brighton, IDS.

MFPED (2002) *Deepening the Understanding of Poverty: Second Poverty Assessment Report,* Uganda Participatory Poverty Assessment Process. Ministry of Finance, Planning and Economic Development, Kampala Uganda. December 2002.

MFPED (2003) *Uganda Poverty Status Report: Achievements and Pointers for the PEAP Revision.* Ministry of Finance, Planning, and Economic Development (MFPED). Kampala, Uganda.

MFPED (2004) An Overview of the National Economy: Government of Uganda. Ministry of Finance, Planning and Economic Development (MFPED). *Discussion Paper 7.*

Micklitz, H.-W. (2000) International Regulations on Health, Safety and the Environment - Trends and Challenges. *Journal of Consumer Policy* 23: 3-24.

Midland Consulting Group (1997) *Beneficiary Assessment of the Agricultural Extension Project cr. 2424-Up,* Ministry of Agriculture, Animal, Industry and Fisheries. Entebbe, Uganda.

Miles, B.M. and Huberman, A.M. (1994) Qualitative Data Analysis: An Expanded Source-Book. Sage Publications, Thousand Oaks.

MISR (2000) *Decentralization: Human Resource Demand Assessment from the Perspective of the District,* Makerere University Institute of Social Research (MISR). Kampala, Uganda.

MOPS (2001) *National Service Delivery Survey,* Ministry of Public Service (MOPS). Kampala Uganda

Moss, G. (1983) *Training Ways: Helping Adults to Learn.* Hasselberg, Wellington.

Mosse, D. (2005) *Cultivating Development: An Ethnography of Aid Policy and Practice.* Pluto, London.

Mubiru, J.B. and Ojacor, F.A. (2001) Agricultural Extension and Education. In: Mukiibi K. J (ed.) *Agriculture in Uganda. Vol I, General information.* Fountain Publishers/CTA/NARO, Kampala. Pp 294-306.

Mukasa, S.B. (2003) Incidence of Viruses and Virus-like Diseases of Sweet Potatoes in Uganda. *Plant Disease* 87 (4): 329-335.

Murphy, P.W. (1990) Creating the Appearance of Consensus in Mende Political Discourse. *American Anthropologist, New Series* 92 (1): 24-41.

Mutandwa, E. and Mpangwa, J.F. (2004) An Assessment of Impact of FFS on IPM Dissemination and Use: Evidence from Smallholder Cotton farmers in the Lowveld area of Zimbabwe. *Journal of Sustainable Development in Africa* 6 (2). www.jsd-africa.com/Jsda/Fall2004/article.htm

Mwagi, G.O., Onyango, C.A., Mureithi, J.G. and Mungani, P.C. (2003) Effectiveness of FFS Approach on Technology Adoption & Empowerment of Farmers: A Case of Farmer groups in Kisii District, Kenya. *Proceedings of the 21st Annual Conference of the Soil Science Society of East Africa. Eldoret, Kenya.*

Nahdy, S. (2004) Uganda: The Uganda National Agricultural Advisory Services (NAADS). In: Rivera, W. and Alex, G. (eds.) Agriculture and Rural Development Discussion paper 8 Extension Reform for Rural Development. *Vol. 1, Decentralized Systems: case studies of international initiatives.* The International Bank for Reconstruction and Development/World Bank.

Narayan, D. (2000) Poverty is Powerlessness and Voicelessness. *Finance & Development, A Quarterly Magazine of the IMF* Vol. 37 No 4 http://imf.org/external/pubs/ft/fandd/2000/12/narayan.htm

NARO (2001) Outreach Partnership Initiative: a Strategy for Decentralized and Institutional Learning. National Agriculture Research Organisation (NARO), Uganda. *Working Paper 1, October 2001*.

NARO (2002) *Addressing the Challenge of Poverty Eradication and Modernization of Agriculture: Improved Technologies by NARO 1192-2002*. National Agricultural Research Organisation (NARO), Uganda naro.go.ug/Information/narodocuments/NARO%20Technology%20Inventory.pdf

Nederlof, E.S. (2006) *Research on Agricultural Research: Towards a Pathway for Client-Oriented Research in West Africa*. Ph.D. Thesis, Wageningen University and University of Ghana, Legon pp 71-97.

Nightingale D.S. and Pindus M.N. (1997) Privatization of Public Social Services. *A Background Paper prepared at Urban Institute for US Department of Labour, Assistant Secretary for Policy, under DOL Contract No. J-9-M-5-0048, # 5*. Accessed on line www.urban.org/publications/407023.html Date accessed 14th September 2006.

North, D.C. (1990) *Institutions, Institutional Change and Economic Performance: Political Economy of Institutions and Decisions*. Cambridge University Press, Cambridge.

Norton, G.A. and Mumford, J.D. (1993) Descriptive Techniques. In: Norton, G. A. and Mumford, J. D. (eds.) *Decision Tools for Pest Management*. CAB International, Wallingford pp 1-21.

Obaa B.B. (2004) *Assessing the Effectiveness of the Contract System of Agricultural Extension with Special Focus on the Enhancement of Demand-driven Processes in Mukono district, Uganda*. M. Sc. Thesis Makerere University. Kampala (U)

Odogola, W., Aluma, J., Sentongo, K.J., Asega, J. and Mugerwa, J. (2003) Farmer Field Schools Approach: a viable methodology for technology development and transfer. *Uganda Journal of Agricultural Sciences*, 8: 427-442.

Okalany, H.D. (1980) *The Pre - Colonial History of the Teso c1490-c1910*, Master in Arts Dissertation, Makerere University. Kampala (U).

Okoth, R.J., Khisa, S.G. and Julianus T (2002) The Journey Towards Self Financed Farmer Field School in East Africa. *Paper Presented at the International Learning Workshop on Farmer Field Schools (FFS): Emerging Issues and Challenges 21-25 October*, Jogyakarta, Indonesia.

Olupot, G., Etiang, J., Aniku, J., Ssali, H. and Nabasirye, M. (2004) Sorghum Yield Response to Kraal Manure with Mineral Fertilizers in Eastern Uganda. In: Bekunda, M.A., Okori, P., Nampala, M.P., Tenywa, J.S., Tusiime, G. and Adipala, E. (eds.) *MUARIK BULLETIN, A research /Journal publication of the Makerere University Agricultural Research Institute, Kabanyolo*, 7: 30-37, MUARIK, Kampala.

Opio-Odong J.M.A. (1989) Organizational Structure and Goal Achievement: The Case of Uganda Extension. In: Semana, A.R,. Opio-Odong, J.A.M. and Zziwa S. (eds.) *Improving Effectiveness of Extension in Uganda*. Department of Agricultural Economics. Makerere University, Kampala.

Opio-Odong J.M.A. (1992) *Designs on the Land: Agricultural research in Uganda, 1980-1990*. African Center for Technology Studies (ACTS). Acts Press, Nairobi.

Ostrom, E. (2005) *Understanding Institutional Diversity*. Princeton University Press, Princeton.

Oxfam and FOWODE (2004) Obusobozi: Enhancing the Entitlements of Subsistence Farmers in Uganda: the Impact of PMA/NAADS on Female Subsistence Farmers. Oxfam GB in Uganda and Forum for Women in Democracy (FOWODE). *A discussion paper by Oxfam GB and FOWODE* pp 53.

References

Oxford Policy Management (2005) *Main Evaluation Report: The Plan for the Modernization of Agriculture*. Main report September 2005.

Ozcatalbas, O., Brumfield, R.D and Ozkan, B. (2004) The Agricultural Information System for Farmers in Turkey. *Information Development 20 (2) 97-102*

Patel, B.K. and Woomer, P.L. (2000) Strengthening Agricultural Education in Africa: The Approach of the Forum for Agricultural Resource Husbandry. *The Journal of Sustainable Agriculture* 16(3): 53-74.

Pawson, R and Tilley, N. (1997) *Realist Evaluation*. Sage Publications, London.

Pawson, R and Tilley, N. (2006) Realist Explanation: Linking Explanatory and Interpretive Accounts in Applied Social Research. *A paper presented in a workshop convened by Technology and Agrarian Development Group Master Class on convergence of explanatory and interpretive approaches within the social and Biological Sciences, 13-14 June 2006, Utrecht and Wageningen University*

PEAP (2004) *Poverty Eradication Action Plan 2004/5-2007/8*. Ministry of Finance, Planning & Economic Development (MFPED). Kampala, Uganda pp 260.

Pedigo, P.L., Hutchins, H.S. and Higley, G.L. (1986) Economic Injury Levels in Theory and in Practice. *Ann. Rev. Entomol.* 31: 341-68.

Pedler, M (Ed.) (1997) Action Learning in Practice (third Edition). Grower Publishing Ltd. Hampshire, England

Place, F., Kariuki, G., Wangala, J., Kristjanson, P., Makauki, A. and Ndubi, J. (2002) Assessing the Factors Underlying Differences in Group Performance Methodological Issues and Empirical Findings from the Highlands of Central Kenya. *Paper presented at the CAPRi workshop on Methodologies for Studying Collective Action, 25 Feb-1st Mar 2002 Nyeri, Kenya.*

PMA SC and MFPED DC 2003 *Guidelines for project/program submission for PMA compliance and Clearance for Funding*. PMA Steering Committee and MFPED Development Committee, Government of Uganda.

PMA (2004) *Annual Report 2003/2004*. Plan for Modernisation of Agriculture (PMA), Kampala, Uganda.

PMA (2005) *Annual Report 2004/2005*. Plan for Modernisation of Agriculture (PMA), Kampala, Uganda.

Pontius, J., Dilts, R. and Bartlett, A. (2002) *From Farmer Field school to Community IPM: Ten years of IPM Training in Asia*. FAO Community IPM Program, Bangkok.

Posner, J.G. and Rudnitsky, N.A. (2001) *Course Design: A Guide to Curriculum Development for teachers* (sixth Edition). Addison Wesley Longman, New York. Chapter 7, PP 151-180.

PRSP (2001) Poverty Reduction Strategic Plan Progress Report: *Uganda Poverty Status Report 2001 Summary*. Ministry of Finance, Planning and Economic Development, Kampala Uganda.

Reason, P. and Bradbury, H. (2001) Introduction: Inquiry and Participation in Search of a World Worthy of Human Aspiration. In: Reason Peter and Bradbury Hilary (eds.) *Handbook of Action Research: participative Inquiry and Practice*. Sage Publications, London pp 1-14.

Richards, P. (1985) *Indigenous Agricultural Revolution: ecology and food production in West Africa*. Hutchinson, London.

Richards, P. (1990) Indigenous Approaches to Rural Development: The Agrarian Populist Tradition in West Africa. In: M. A. Alteri and S. B. Hecht (eds.) *Agroecology and Small Farm Development.* CRC Press, Boca Raton FA pp 105-111.

Richards, P. (1993) Cultivation or Performance: In: Hobert, M. (Ed.), *An Anthropological Critique of Development: The Growth of Ignorance.* Routledge, London pp 61-78.

Archibald, S. and Richards, P (2002) Seeds and Rights: New Approaches to Post-war Agricultural Rehabilitation in Sierra Leone. *Disasters* 26 (4): 356-367.

Richards, P. (2006) Against Discursive Participation: Authority and Performance in African Rural Technology Development. *Paper presented at European Association for the Study of Science and Technology (EASST) Conference, 23rd-26th August 2006;* University of Lausanne, Switzerland.

Riesenberg, L.E. (1989) Farmers' Preferences for Methods of Receiving Information on New or Innovative Farming Practices. *Journal of Agricultural Education, Fall,* pp 7-13 http://pubs.aged.tamu.edu/jae/pdf/Vol30/30-03-07.pdf

Rivera, W. and Alex, G. (Eds) (2004) Volume 3. Demand-Driven Approaches to Agriculture Extension: Case Studies of International Initiatives. *Agriculture and Rural Development Discussion paper 10 Extension Reform and Rural Development.* The World Bank, Washington D. C.

Rivera, W. and Alex, G. (2005) Extension Reform: The Challenges Ahead. *Proceedings of the 21st Annual Conference, AIAEE.* San Antonio, TX pp 260-271.

Rola, A.C., Jamias, S.B. and Quizon, J.B. (2002) Do FFS Graduates Retain and Share What they Learn? An Investigation in Iloilo, Philippines. *Journal of International Agricultural Extension and Education* 9 (1): 65-75.

Röling, N.G. and Engel, P.G.H. (1990) Information Technology from a Knowledge Systems Perspective: Concepts and Issues, Knowledge in Society. *The International Journal of Knowledge Transfer* 3 (3): 6-18.

Rollins, J.T. and Golden, K. (1994) A Proprietary Information Dissemination and Education System, *Journal of Agricultural Education* 35 (2):37-42.

Rwabwoogo, O.M. (2002) *Uganda Districts: Information Handbook.* Fountain Publishers, Kampala, Uganda.

Sanchez, D. and Cash, K. (2003) *Reducing Poverty or Repeating Mistakes? A Civil Society Critique of Poverty Reduction Strategy Papers.* Church of Sweden Aid, Diakonia, Save the Children Sweden, and the Swedish Jubilee Network, Brussels, Eurodad.

Scoones, I. and Thompson, J. (1994) Beyond Farmer First: Rural People's Knowledge, Agricultural Research and Extension Practice. Intermediate Technology Publications, London.

Selener, D. (1997) Participatory Action Research and Social Change. Cornel university, Ithaca NY

Semana A.R., Opio-Odong, J.M.A and Zziwa, S. (eds.) 1989 Improving Effectiveness of Extension Services in Uganda. Department of Agricultural Economics, Makerere University. Kampala (U).

Semana, A.R. (1989) Agricultural Extension Approaches and Methods in Uganda: General Aspects. In: Semana, A.R., Opio-Odong, J.M.A. and Zziwa S. (eds.) *Improving Effectiveness of Extension services in Uganda.* Department of Agricultural Economics, Makerere University.

Semana, A.R. (2002) Agricultural Extension Services at Cross-roads: Past, Present, Dilemma and Possible solutions for future in Uganda. *Proceedings of CODESRIA-IFS Sustainable Agriculture Initiative Workshop in Kampala, Uganda, 15-16, 2002 December.*

References

Shucksmith, M (2000) Endogenous Development, Social Capital and Social Inclusion: perspectives from LEADER in the UK. *Sociologia Ruralis* 40 (2): 208-218.

Sibyetekerwa, P.E.K. (1989) A review of the Strengths and Weaknesses of Agricultural Extension Services in Uganda. In: Semana, A.R., Opio-Odong, J.M.A. and Zziwa S. (eds.) *Improving Effectiveness of Extension services in Uganda*. Department of Agricultural Economics, Makerere University.

Silverman, D. (2001) *Interpreting Qualitative Data: Methods for Analyzing Talk, Text and Interaction* (second edition). Sage Publications Ltd, London.

Simmons, R and Birchall, J. (2005) A Joined-up Approach to user Participation in Public services: Strengthening the "Participation chain" *Social Policy & Administration* 39 (3): 260-283.

Simpson, B.M. and Owen, M. (2002) Farmer Field Schools and the future of agricultural Extension in Africa. *Proceedings of the 18th Annual Conference AIAEE, Durban, South Africa*, pp 405-412.

Simsons, T., Pelled, H.L. and Smith, A.K. (1999) Making Use of Difference: Diversity, Debate and Decision Comprehensiveness in Top Management Teams. *Academy of Management Journal* 42 (6): 662-673.

SOCADIDO (2001) *Activity Report (2001)* Soroti Catholic Diocese Integrated Development Organization (SOCADIDO) Soroti, Uganda.

Sorby, K., Gerd, F. and Eija, P. (2003) Integrated Pest Management in Development: Review of Trends and Implementation Strategies. *Agriculture and Rural development Working Paper 5*. Washington, D.C.: World Bank.

Stathers, T., Olupot, M., Khisa, G., Kapinga., R. and Mwanga, R. (2006) *Final technical Report of the Expansion of Sustainable Sweetpotato production and Post harvest management through farmer field schools in East Africa and sharing of lessons learnt project*. Natural Resources Institute, Chatham.

Stern, V.M. (1973) Economic Thresholds. *Ann. Rev. Entomol*. 18: 259-80.

Stringfellow, R., Coulter, J., Lucey, T., McKone, C and Hussain, A. (1997). Improving the Access of small-holders to Agricultural Services in Sub-Saharan Africa: Farmer Copperation and the Role of the Donor Community. *ODI Natural Resource Perspectives, Number 20*, June 1997.

Sutherland, A., Martin, A. Smith, R.D. (2001) *Dimensions of Participation: Experiences, lessons and Tips from Agricultural Practitioners in sub-Saharan Africa*. Natural Resource Institute, Chatham Maritime.

Swenson C.L. (1980) *Theories of Learning: Traditional Perspectives, Contemporary Developments*. Wadsworth, Belmont.

Tanner, D. and Tanner, L. (1995) Curriculum Development: Theory and Practice (third edition). Prentice Hall, New Jersey.

Tesfaye, A., Jemal, I., Ferede, S. and Curran, M.M. (2005) Technology Transfer Pathways and Livelihood Impact Indicators in central Ethiopia. *Tropica Animal Health and Production* 37 (Suppl. 1): 101-122.

Thiele, G., van de Fliert, E. and Campilan, D. (2001) What Happened to Participatory Research at the International Potato Center? *Agriculture and Human Values* 18: 429-446.

Tripp, R., Wijeratne, M. and Piyadasa, V.H. (2005) What Should We Expect from Farmer Field Schools? A Srilanka Case Study. *World Development* 33 (10): 1705-1720.

Tuckman, B.W (1965) Developmental Sequences in Small Groups. *Psychological Bulletin* 63, 384-399

Tuckman, B.W. and Jensen, M.A. (1977) Stages of Small Group Development Revisited. *Group and Organisational Studies* 2: 419-427.

Turrall, T., Mulhall, A., Rees, D., Okwadi, J., Emerot, J. and Omadi, R. (2002) *Understanding the Communication Context in Teso and Lango Farming Systems: The Agricultural Information Scoping Study. Executive Summary (Revised March 2002)* NARO/DFID Client-Oriented Agricultural Research and Dissemination Project 17pp.

UBOS (2002) *Uganda Population and Housing Census, main report*, Uganda Bureau of Statistics (UBOS). Entebbe, Uganda.

UBOS (2006) *Statistical Abstract* Uganda Bureau of Statistics (UBOS). Entebbe, Uganda.

UNDP (2001a) *Review of the Poverty Reduction Strategy Paper (PRSP)*. United Nations Development Program (UNDP), New York.

UNPD (2001b) *Making new Technologies Work for Human Development* United Nations Development Program (UNDP). Oxford University Press, Oxford.

UPPAP (2002) *Second Participatory Poverty Assessment Report: Deepening the Understanding of Poverty*. Ministry of Finance and Economic Planning (MFPED) Kampala, Uganda.

UPPAR (2000) *Uganda Participatory Poverty Assessment Report*. Ministry of Finance, Planning and Economic Development, Kampala, Uganda.

Van de Fliert, E. (1993) Integrated Pest Management: Farmer Field Schools Generate Sustainable Practices. A Case Study in Central Java Evaluating IPM Training. *Wageningen Agricultural University Papers 93-3 and Ph.D. Thesis*. The Netherlands.

Van de Fliert, E. (2000) Stepping Stones and Stumbling Blocks in Capacity Development of Sweet Potato ICM farmer Field School Facilitators. *Paper presented at the UPWARD Conference "Capacity Development for Participatory Research" Beijing 19-22 September 2000.*

Van de Fliert, E., Thiele, G., Campilan, D., Ortiz, O., Orrego, R., Olanya, M. and Sherwood, S. (2002) Development and Linkages of Farmer Field School and other Platforms for Participatory Research and Learning. *Paper presented at the International Learning Workshop on Farmer Field School (FFS): Emerging Issues and Challenges, 21-25 October, 2002. Yogyakarta, Indonesia.*

Van Huis, A. and Meerman, F. (1997) Can we Make IPM Work for Resource Poor Farmers in Sub-Saharan Africa? *International Journal of Pest Management* 43 (4): 313-320.

Van Mele, P., Salahuddin, A. and Magor, P.N. (2005) People and Pro-poor Innovation Systems. In: Van Mele, P., Salahuddin, A. And Magor, P. N (eds.), *Innovations in Rural Extension: case studies from Bangladesh*. CABI Publishing, Wallingford, UK pp 257-296.

Walaga, C., Egulu, B., Bekunda, M. and Ebanyat, P. (2000) Impact of Policy on Soil Fertility Managementin Uganda. In: Hilhorst, T. and Muchena, F. M. (eds.) *Nutrients on the Move: Soil Fertility Dynamics in African Farming Systems*. International Institute for Environment and Development (IIED), London pp 29-44.

WICCE (2002) Documentation of Teso Women's Experiences of Armed Conflict 1987-2001. *An Isis-WICCE Research Report, Part One*, Isis-Women's International Cross Cultural Exchange. Kampala, Uganda pp77.

Webster, B.J., Ogot B.A. and Chretieng, J.P. (1992) The Great Lakes' Region 1500-1800 Ogot, B. A (ed.) *General History of Africa. V: Africa from the Sixteenth to the Eighteenth Century*. Hernemann, UNESCO pp 776-827.

References

Whitaker, P. (1995) *Managing to Learn: Aspects of Reflective and Experiential Learning in Schools.* Cassell, London.

Winarto, T.Y. (2004) *Seeds of Knowledge: The Beginning of Integrated Pest Management in Java.* Yale University Southeast Asia Studies, New Haven.

Wortman, C.S. and Kayizzi, C.K. (1998) Nutrient Balances and Expected Effects of Alternative Practices in Farming Systems of Uganda. *Agriculture, Ecosystems & Environment* 71: 115-129.

Yin, R.K. (2003) *Case Study Research: Design and Methods* (Third Edition). Sage Publications, Thousand Oaks.

Zake, J.Y.K. (1993) A Review of Soil Degradation and Research on Soil Management in Uganda. In: Pauw A. de (Ed.), *The Management of Acid Soils and Land Development. Network document 13,* IBSRAM, Bangkok.

About the Author

Prossy Isubikalu was born on 15th January 1973 in Iganga district, Uganda. She joined Makerere University in 1992 where she graduated with a B.Sc. Agriculture and M.Sc. Agricultural extension in 1996 and 1999 respectively. In her M.Sc. work, Prossy looked at "Understanding farmer knowledge of cowpea production and pest management: case studies from eastern Uganda". After completing the second degree, she was employed by local government of Iganga district as an agricultural extension officer in charge Bulamagi sub-county in June 1999. In August, 2001 she was employed as an assistant lecturer in the department of Agricultural Extension Education, Faculty of Agriculture, Makerere University.

In February 2002, Prossy commenced her Ph.D. studies in Wageningen University under the Technology and Agrarian Chairgroup. While at wageningen, she attended a workshop orgnised by the Participation and Up-scaling programme (PAU) that triggered her interest in facilitation. With this knowledge and skill, Prossy has been able to design, facilitate and document a number of workshops, mainly at local level. Her areas of interest include community mobilisation and development, facilitation, understanding and analysing participation, personal development and action research.

Printed in the United States
by Baker & Taylor Publisher Services